Seismic Inversion

Other Books in the Series
Petroleum Geostatistics

Coming Soon
Naturally Fractured Reservoir Characterization

Seismic Inversion

Mrinal K. Sen
U. of Texas at Austin
John A. and Katherine G. Jackson School of Geosciences

Society of Petroleum Engineers

ISBN 978-1-55563-110-9

06 07 08 09 10 11 12 13 / 9 8 7 6 5 4 3 2 1

Society of Petroleum Engineers
222 Palisades Creek Drive
Richardson, TX 75080-2040 USA

http://www.spe.org/store
service@spe.org
1.972.952.9393

Preface

During the last decade, it has become a routine practice to employ seismic inversion for rock property estimation. There are essentially two principal uses of seismic data, namely, obtaining an image of the earth's subsurface and estimating elastic properties of the subsurface rock layers. While imaging is focused on obtaining the spatial locations of the interfaces, seismic inversion is aimed at converting the interface information to interval information. Simply stated, the goal of inversion is to estimate earth model parameters from remotely sensed geophysical data. Estimation of the slowly varying component of the velocity field for seismic imaging is indeed an inverse problem. However, most commonly the term *seismic inversion* is used for estimating interval elastic properties from seismic data. The results from such inversions are used for reservoir model-building. For example, an inverted impedance volume can be tied to well logs and directly related to reservoir properties. Such a volume is also used to interpolate well logs between sparsely located wells. These results offer new insights in seismic reservoir characterization and 4D or time-lapse seismic studies.

The primary goal of this primer is to provide a review of the seismic inversion methods. The target audiences for this primer are nonspecialists from different branches of science and engineering. Keeping this goal in mind, I provide a summary of the commonly used seismic methods and inverse theory, and describe in detail different inversion methods for estimating layer properties. There are two important aspects of seismic inversion: one is the development of algorithms, and the other is the interpretation of the results in terms of rock parameters, including lithology estimation and detection of hydrocarbon. Neither of these is a trivial task. Seismic inversion results in nonunique estimates of elastic properties and is an ill-posed nonlinear problem. Including prior information and developing robust techniques are the fundamental goals in designing an inversion algorithm. The nonuniqueness aspect of seismic inversion gets further complicated by the fact that the elastic parameters—such as compressional and shear-wave velocities and impedance—are affected by a large number of environmental parameters. Often, there is a large degree of overlap between these properties for different rock types. Interpretation of the results from seismic inversion in terms of rock types, lithology, and occurrence of hydrocarbon is the most important aspect of seismic inversion. Sound knowledge of geology is essential to obtaining meaningful results from seismic inversion. I highlight several of these aspects in this text. My description of seismic inversion is not complete by any means, because better and powerful techniques are being developed continually. I have attempted to provide a concise view of the "state of the art" in seismic inversion.

Sincere thanks to Hossein Kazemi, Scott Tinker, and the SPE Publications Committee for inviting me to write the primer; to Paul Stoffa for continued collaboration on seismic imaging and inversion research over the years; to Janet Everett for formatting the manuscript; to Jim Jennings, Jay Pulliam, Tiangkong Hong, Carl Sondergeld, and Raghu Chunduru for reviewing the manuscript; and to Dhananjay Kumar, Vik Sen, I. Ahmed, R.K. Shaw, and I.G. Roy for their help with the preparation of several figures. Several of my graduate students and post-doctoral fellows participated in seismic inversion research; their results have been freely raided. Sincere thanks to Jack Dvorkin for generously sharing originals of two of his published figures. Figs. 4.3 and 4.8 were generated with a commercial software, Jason Workbench. In addition, I have made use of several published results. Because of the limited length of the primer, I am unable to cite many important articles. Finally, I thank my wife Alo and my children, Amrita and Ayon, for their love, patience, and cooperation.

In memory of my father,
Ram Prasad Sen, a loving teacher and a caring social worker

Contents

Chapter 1

Seismic Exploration Fundamentals

Opportunities for direct observation of subsurface rocks are limited to outcrops, mines, tunnels, and caves, which together offer only a glimpse of the inner Earth. Much more information on subsurface geology is obtained from data collected when drilling for hydrocarbons and groundwater. An even greater degree of knowledge is gained from remotely sensed geophysical data, from which the derivation of subsurface geology is rather indirect. A comprehensive description of subsurface rock properties is obtained by the judicious use of multiple data sets. Of all the geo-exploration datasets, geophysical data are the most inclusive; their information content ranges from near-surface layers to the inner core of the Earth. The detection capabilities and resolution of the geophysical methods are also wide-ranging. Of all the geophysical exploration methods, the seismic method is unequivocally the most important, primarily because it is capable of detecting large global-scale to small-scale features. It is also the most expensive of the geophysical methods and is principally employed in exploration for oil and gas and mapping bedrock characters. Several introductory and advanced textbooks (e.g., Sheriff and Geldart 1995; Telford *et al.* 1990) describe the principles of seismic exploration. In this chapter, we will review the fundamental concepts employed in seismic exploration.

Simply stated, seismic methods involve estimation of shapes and physical properties of Earth's subsurface layers from the echoes of sound waves that are propagated through the subsurface. The basic ideas of seismic exploration are borrowed from earthquake seismology. Earthquakes are usually caused when subsurface rock suddenly breaks; that is, an earthquake is a movement or shaking of the Earth caused by movement of the rock layers along a fault plane. Such a rupture in a fault causes subsurface vibrations that propagate as seismic waves **(Fig. 1.1)**. Seismic instruments, called *seismographs*, placed on the Earth's surface record the ground motion (particle displacement, velocity, or acceleration) as a function of time. Such recorded time histories of ground motion are called *seismograms*, and they carry information about the variations in subsurface structure and properties sampled by the seismic waves during propagation from the source to the recording instrument. A complete suite of such records is analyzed to infer properties of the Earth's interior. Most of what we know today about the Earth's deep interior has been derived from seismology. However, obtaining detailed maps of the Earth's interior depends on the location of the

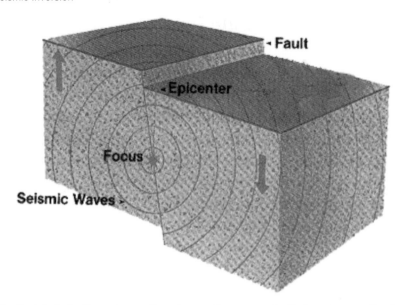

Fig. 1.1—Model of a fault causing earthquakes: a slip across a fault plane causes vibrations that propagate throughout the medium.

earthquake sources and the placement of the recording instruments. Earthquakes take place in regions where the subsurface conditions are favorable for rock failures to occur and are usually restricted to very few regions of the Earth, such as along tectonic plate boundaries and faults. They are recorded only at places where the instruments are placed. An earthquake seismology experiment can be termed a passive experiment in that the instruments listen to ground vibrations when an earthquake occurs.

In contrast to a passive experiment, a seismic exploration is an active experiment, which is carefully planned. Ground is made to vibrate, and recording instruments are placed at locations that the Earth scientists and engineers are interested in studying. The primary focus of seismic exploration is restricted to the upper few kilometers of the crust, with the goal of obtaining detailed pictures of the subsurface that can only be derived from an active experiment. We make use of artificial sources such as explosions to cause ground vibrations, and hope to achieve continuous coverage, in which the seismic response from subsurface layers of increasing depth is sampled along profiles of seismic recorders usually placed on the surface of the Earth, but occasionally placed in boreholes. It is important to note that most seismic methods map geological structures favorable for the accumulation of hydrocarbons, rather than hydrocarbons directly. The seismic data alone cannot satisfy our exploration objectives; only when supplemented with other data sets can a unique interpretation of seismic data be made. Modern methods make use of seismic amplitudes (relative ground motion) and advanced techniques of inversion to infer properties of subsurface rocks and possibly their hydrocarbon content.

1.1 Seismic Experiments
1.1.1 Survey Geometry. Seismic surveys are conducted both on land and in the ocean. Airborne surveys are also possible, but have found few applications. In a land survey **(Fig. 1.2)**,

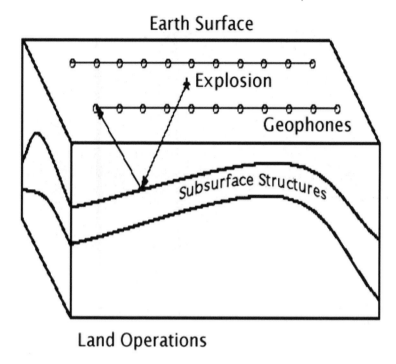

Fig. 1.2—Layout of a land 3D seismic survey: explosions from dynamite generate vibrations and the returned echoes are recorded on geophones spatially distributed around the source.

a source is excited at a point, and the recorders are placed at several locations on the ground. In practice, the sources are usually buried, and the receivers (geophones) are well coupled to the ground. For a fixed array of receivers, the source location is varied, and then the receiver arrays are moved, and the process is repeated. This results in a broad range of data coverage, causing an increase in "signal-to-noise" ratio. In a typical marine survey **(Fig. 1.3)**, a ship tows an array of receivers while shooting air guns at regular intervals as it moves. Modern 3D experiments make use of multiple cables (six or more), each of which can be as long as 8 km and can carry hundreds of recorders. Marine surveys occasionally record seismic waves using instruments placed on the seafloor **(Fig. 1.4)**; they include ocean-bottom seismometers (OBS) and ocean-bottom cables (OBC). Vertical seismic profile (VSP) surveys place recorders in boreholes while shots are fired at the surface and vice versa.

1.1.2 Energy Sources. Seismic sources play an important role in our ability to detect subsurface features. Ideally we would like to be able to inject sufficient energy that is detectable after it has traveled though the exploration target zones. Sources should also be designed so that echoes from different rock boundaries are distinguishable. Land surveys typically make use of dynamite in small holes 6–30 m in depth. One variant of land seismic sources is a *Vibroseis,* a device that uses a sweep with frequencies varying between 12 and 60 Hz. Special processing of the recorded data is required to retrieve information from such records. The air guns used as sources in marine surveys discharge air under high pressure (approximately 2000 psi) into the water.

Sea Surface

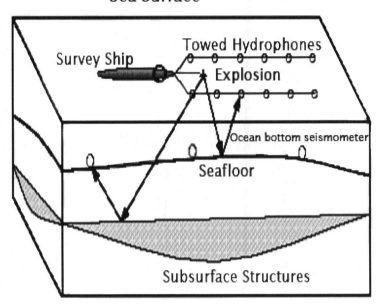

Fig. 1.3—A marine 3D seismic survey: a ship tows streamers containing hydrophones, and waves are generated by air guns.

1.1.3 Source-Time Function: The Wavelet. As dynamite is exploded or an air gun is fired, the energy level varies with time; the measure of energy level as a function of time at the source location is termed a source-time function or a wavelet. Measurement of a wavelet at the source level is not often done, and therefore, the wavelet needs to be unraveled from recorded seismic data. Obviously, the wavelet plays an important role in seismic data analysis and is crucial to obtaining high-resolution images. Ideally, a delta function or a spike-like source is desired, because only such a wavelet would enable us to identify individual layer boundaries. In practice, however, such an ideal source-time function is impossible to achieve. Even if it was possible to generate a delta-function-like wavelet at the source location, such a shape is not retained, because the Earth acts as a filter in which high frequencies are attenuated as the energy propagates through the earth. In other words, the shape of the wavelet changes with time. For all practical purposes, a wavelet is assumed to be stationary and band-limited; often an average wavelet is estimated from the seismic data.

In seismic data processing, the wavelet is usually removed, and a simpler, so-called zero-phase wavelet is convolved. The most common zero-phase wavelet is a *Ricker wavelet* (Ricker 1953) described by the equation

$$f(t) = (1 - 2\pi^2 \omega_{max}^2 t^2)\exp(-\pi^2 \omega_{max}^2 t^2), \qquad \ldots\ldots\ldots\ldots\ldots\ldots\ldots\ldots\ldots(1.1)$$

where $f(t)$ is the amplitude of the wavelet at time t and ω_{max} is the peak frequency of the wavelet. **Fig. 1.5** displays Ricker wavelets for a series of peak frequencies.

Fig. 1.4—Typical academic ocean bottom seismometer equipment (OBS); the picture shows deployment of the OBS.

1.1.4 Measured Variables. The instruments used in land seismic experiments are called *geophones;* they measure ground displacements (or velocity or acceleration, which are all vector quantities) as a function of time at locations at which they are buried. Similarly, marine experiments make use of hydrophones that record variations in pressure (a scalar quantity) in the water at the hydrophone locations as a function of time. OBS and OBC surveys employ both hydrophone and three-component geophones. The hydrophone component floats in water, recording pressure variations, while the geophones anchored to the seafloor record three components (one vertical and two horizontal) of ground motion. Records of such variations in pressure and ground displacements, both as a function of time and spatial locations of the phones, are analyzed further to obtain images and derive elastic properties of the subsurface layers.

1.2 Reflection

When dynamite is exploded in a borehole or an air gun is fired in the water, a disturbance is created in the surrounding medium, which propagates through the medium as waves. As these waves encounter a contact surface between two rock layers (say, sandstone and

Ricker Wavelet

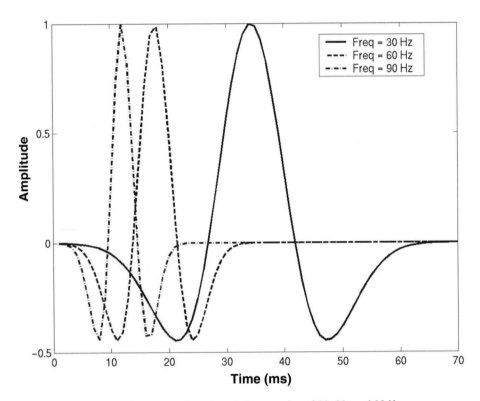

Fig. 1.5—Ricker wavelets at peak frequencies of 30, 60, and 90 Hz.

shale), part of the energy is reflected back, and the remaining part is transmitted into the next layer underneath. The reflected energy propagates upward and is recorded at the receiver location when it arrives there. Such a phenomenon can be explained well using ray paths—a concept that is introduced in elementary optics. For example, a stack of flat-lying layers with varying rock properties is shown in **Fig. 1.6a**. The ray diagram shown in Fig. 1.6a depicts the paths traversed by the waves during propagation from the source to the receiver via the reflecting medium. *Snell's law* dictates the reflected and transmitted ray paths. The seismograms corresponding to a set of receivers shown in Fig. 1.6a are displayed in Fig. 1.6b; distinct reflections from the layer boundaries are marked. Reflections from thicker layers are better separated than those from thinner layers, and as the separation between the source and the receiver increases, the waves take progressively longer to arrive at the receiver. For more complex geologic models, the ray paths and the corresponding seismograms are more complicated than the simple layered (vertically varying or 1D) Earth models, as shown in **Fig. 1.7**.

1.3 Seismic Wave Velocities
A detailed and more rigorous description of seismic wave propagation is outlined in Chapter 2. Here we will make use of key concepts to explain some of the fundamental features

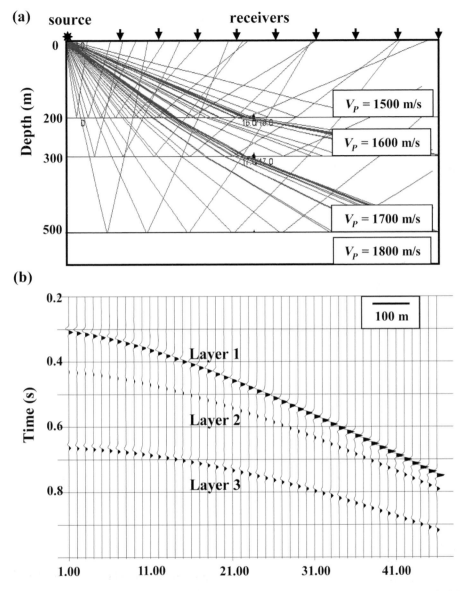

Fig. 1.6—(a) A stack of horizontal layers with the corresponding ray-diagrams (V_S is half of V_P and density is constant at 1.0 g/cm^3); (b) seismograms corresponding to the model shown in (a).

observed in the seismograms. The Earth is assumed to be an elastic medium. In such a medium, a rock layer is characterized by two elastic parameters called the *Lamé parameters*, represented by symbols λ and μ, as well as the density, ρ. The SI unit of λ and μ is N/m^2. Two types of waves are often found to propagate through the Earth. They are called *Primary*, or *P-waves*, and *Shear*, or *S-waves*. The two waves have distinct propagation behavior and particle motion. For example, particle motion for the P-wave is always in the direction of propagation, while that of the S-wave is perpendicular to the direction of propa-

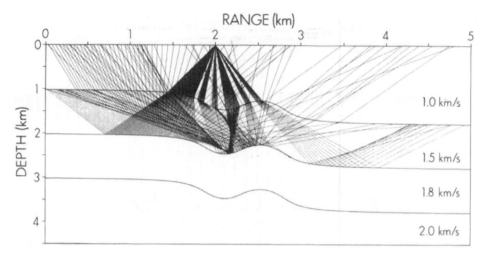

Fig. 1.7a—A laterally varying Earth model with the corresponding ray paths.

gation. The seismograms shown in Figs. 1.6 and 1.7 display P-wave recordings only. As noted in the previous section, the seismic wave velocity that is the characteristic of the medium through which the waves propagate determines the time of arrival of seismic energy. The velocities of propagation of P- and S-waves are given by the expressions

$$\alpha = \sqrt{\frac{\lambda + 2\mu}{\rho}} \; ; \quad \beta = \sqrt{\frac{\mu}{\rho}} \, , \quad \dots\dots\dots\dots\dots\dots\dots\dots\dots\dots\dots\dots\dots\dots (1.2)$$

where α and β represent P- and S-wave velocities, respectively.

Ideally, if we are able to derive P- and S-wave velocities of different rock layers from the arrival times of the seismic waves, we should be able to relate those to rock types and possibly predict hydrocarbon zones. Unfortunately, this is easier said than done. First, sedimentary rocks such as sandstone are not made up of simple elastic layers; they are porous, and the pores contain fluids of different types, such as water, oil, and gas. Second, laboratory and in-situ measurements of seismic wave velocities show that there exists significant overlap in seismic wave velocities between sandstones and shales, two common rock types encountered in hydrocarbon-bearing formations.

In sedimentary rocks, a more rigorous poroelastic model that takes into account petrophysical parameters such as porosity, fluid saturation, etc., should be used to model seismic wave behavior. In practice, however, empirical relations are used to relate these petrophysical parameters to the elastic moduli, and an elastic model is used for simulation of wave propagation. Detailed descriptions of this subject can be found in Sheriff and Geldart (1995) and Mavko *et al.* (1998).

Factors affecting compressional and shear-wave velocities of sedimentary rocks are *lithology, porosity, density, lithification* or *cementation, pressure* or *depth of burial*, fractures, and *fluid content* or *fluid saturation*. Of all these parameters, porosity is the most important factor for determining velocity. An approximate formula known as *Wyllie's time-average equation* (Wyllie *et al.* 1958), is commonly used to compute velocity. This equation is given by

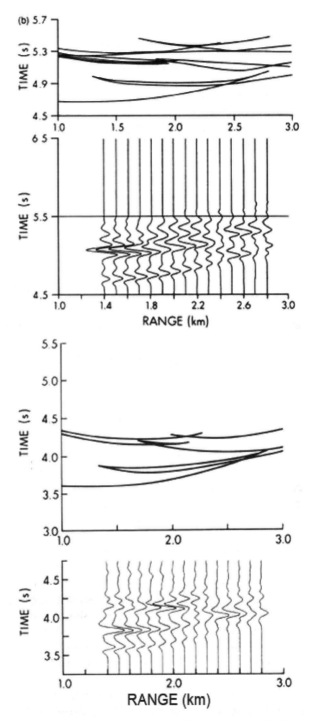

Fig. 1.7b—Corresponding reflection response from the reflector. Only P-wave primary reflections are shown. Note the complexity of the ray paths and the seismograms. The top two panels show travel time and the seismograms corresponding to reflection from the shallow layer, whereas the same for the second layer are shown in the bottom two panels.

$$\frac{1}{v_{\text{bulk}}} = \frac{\phi}{v_{\text{fluid}}} + \frac{(1 - \phi)}{v_{\text{matrix}}}, \quad \dots \dots \dots \dots \dots \dots \dots \dots \dots \dots (1.3)$$

where ϕ is the porosity, v_{fluid} is the fluid velocity and v_{matrix} is the velocity of the rock matrix. Thus, given the porosity and the velocities of the fluid and the rock matrix, the effective bulk slowness (reciprocal of velocity) can be computed by the weighted average given by the above equation.

Lithification or cementation is the degree to which grains in a sedimentary rock are cemented together by post-depositional chemical processes. This has a strong influence on the elastic moduli. By filling pore spaces with minerals of higher density than fluids, the bulk density is increased. A combination of reduction in porosity and cementation causes the observed increase in velocity with depth of burial and age.

The compressional wave (P-wave) velocity is strongly dependent on effective stress. For a rock buried in the Earth, the confining pressure is the pressure of the overlying rock column. The pore water pressure may be equal to the hydrostatic pressure if there is connected porosity to the surface, or it may be greater or smaller than the hydrostatic pressure. The effective pressure is the difference between the confining and pore pressures. In general, the velocity increases with an increase in confining pressure and then it attains a terminal velocity when the confining pressure is high, due possibly to crack closure. If the pore pressure exceeds the hydrostatic, the effective stress is reduced, causing a decrease in velocity. Such overpressured zones can be detected in a sedimentary sequence by their anomalously low velocities.

Pore spaces within rocks are almost always filled with fluids such as saline water, gas, or oil. Saturation of pore fluid has a significant effect on the P-wave velocity, which is generally found to increase with increasing fluid saturation. When pore-water is replaced with oil or gas, changes in bulk density and elastic parameters occur, causing changes in velocities. Replacement of water or oil with gas does not change the shear modulus appreciably but reduces the density significantly. Thus, the shear-wave velocity changes by a small amount, but the P-wave velocity drops significantly—a property used widely in amplitude analysis for direct detection of hydrocarbons. The effects of all these parameters have been discussed in great detail in Tatham and McCormack (1998).

1.4 Rock Physics/Fluid Substitution Model

In addition to obtaining images of a hydrocarbon reservoir, one of the primary goals of seismic exploration is to obtain estimates of porosity and fluid saturation within the reservoir. As discussed in the previous section, much seismic analysis is done with elastic wave theory. A rock physics/fluid substitution model allows one to establish a quantitative link between fluid-flow petrophysical parameters and effective elastic properties. Several models exist for such transformation, of which the *Biot-Gassmann equations* (Biot 1956; Gassmann 1951) are the most popular. The basic equations for P-wave and S-wave velocities in isotropic, elastic nonporous media are given in terms of the bulk and shear moduli by

$$\alpha = \sqrt{\frac{K + \frac{4}{3}\mu}{\rho}}, \quad \dots \dots \dots \dots \dots \dots \dots \dots \dots \dots (1.4)$$

and

$$\beta = \sqrt{\frac{\mu}{\rho}}, \quad \dots\dots\dots\dots\dots\dots\dots\dots\dots\dots\dots\dots\dots\dots \text{(1.5)}$$

where K is the rock's bulk modulus (in turn a function of porosity, ϕ, of water saturation, S_w, and of the mechanical properties of the rock's solid skeleton), μ is the shear modulus, and ρ is the bulk density. The bulk density can be written as

$$\rho = \phi S_W \rho_W + \phi(1 - S_W)\rho_o + (1 - \phi)\rho_S, \quad \dots\dots\dots\dots\dots\dots\dots\dots \text{(1.6)}$$

where ρ_w, ρ_o, and ρ_s are the densities of water, oil, and of the rock's solid skeleton, respectively. For simplicity, if we assume a fluid model consisting of only water and oil, the rock's bulk modulus K of the saturated rock can be estimated from *Gassmann's equation* (Nur and Wang 1998), namely,

$$K = \left(K_b + \frac{4}{3}\mu\right) + \frac{\left(1 - \frac{K_b}{K_s}\right)^2}{\left(1 - \phi - \frac{K_b}{K_s}\right)\left(\frac{1}{K_s}\right) + \frac{\phi}{K_f}}, \quad \dots\dots\dots\dots\dots \text{(1.7)}$$

where subscripts b, s, and f stand for bulk (dry, porous frame property), solid and fluid, respectively. Note that Gassmann's equation makes these assumptions:

- The rock is homogeneous and isotropic.
- The pore space is completely connected.
- Wave propagation is restricted to low frequencies (<100 Hz) and the fluid is nonreactive.

The porosity, elastic properties, and water saturation relationships summarized by Eq. 1.4 through 1.7 quantitatively link the wave equation with the two-phase fluid-flow equation. Despite the assumptions made in Gassmann's equation, these relationships have been found to model several observations of seismic wave velocities and are, therefore, used frequently (Smith *et al.* 2003).

1.5 Seismic Resolution

One of the questions most commonly encountered by practicing geophysicists is "what is the level of detail one can decipher from seismic data?" Seismic resolution is the ability to distinguish separate features (i.e., "what should be the minimum distance between two features so that the two can be defined separately in the seismic data?"). To answer this question, we must consider both vertical and horizontal resolutions. The estimate of the vertical resolution is given by the so-called *Rayleigh criterion*, which states that in order for two reflecting interfaces to have a distinct reflection signature, they should be at least 1/4 of the wavelength apart. For layers thinner than a 1/4 wavelength, we rely on the wave amplitude. Recall that the velocity is given by the following relation:

Velocity = frequency × wavelength.

Thus the frequency of the wavelet determines the vertical resolution of seismic data. That is, the higher the frequency, the finer the details of the subsurface features that can be discerned from the seismic data. Obviously, the resolution is also controlled by the velocity, which varies laterally as well as with depth. The velocity variation with depth is generally more pronounced than the lateral variations; typically, the seismic wave velocity increases with depth. For example, in the upper 10 km of the Earth, for an average P-wave velocity of 1000 m/s and for a frequency of 100 Hz, the resulting wavelength is 10 m with a vertical resolution of about 2.5 m. Note, however, that with increasing depth, high frequencies are attenuated and velocity increases. Thus, at a depth of 5 km for a velocity of 5 km/s and a frequency of 20 Hz, we have a wavelength of 250 m.

Horizontal resolution refers to how closely two reflecting points should be situated horizontally so that they can be recognized as two separate points. The answer to this question is given by the width of the so-called *First Fresnel zone*. Like the vertical resolution, the horizontal resolution also depends on the frequency and the velocity, which determine the width of the Fresnel zone. To understand this, recall that a reflection is not the energy reflected from just one point. The width of the first Fresnel zone is the reflecting zone in the subsurface insonified by the first quarter of a wavelength. If the wavelength is long, the zone from which the reflected returns come is larger and the resolution is lower. For example, for a depth of 3 km and a velocity of 3 km/s, the Fresnel zone radius ranges from 300 to 470 m for frequencies of 50 to 20 Hz.

1.6 Seismic Processing

Seismic data are used for obtaining maps of the subsurface and thereby calculating detailed estimates of petrophysical properties of subsurface rocks for use in reservoir characterization. To achieve the first goal, seismic traces are lined up to obtain a section that is expected to resemble subsurface structure. For example, consider a very simple scenario in which we have a stack of horizontal layers and a seismic experiment consisting of coincident source-receiver pairs. As the coincident source-receiver pair is moved along the surface, a seismic trace is recorded at each surface location; a reflection from each interface arrives at a time corresponding to the two-way-time, or normal incident reflection time, from this interface. If the velocity is known, the reflection time series can be converted to depth. When such depth-converted traces are displayed as a profile, called a *zero-offset section* (**Fig. 1.8**), they represent a geologic structure. Unfortunately, in practice this is not an easy task, because the Earth is laterally heterogeneous (it is the heterogeneity that we are interested in), and it is almost impossible to have coincident source-receiver pairs. Even if we were able to design a coincident source-receiver experiment, the reflections recorded by the seismograms will rarely correspond to points from the subsurface vertically below the source-receiver.

Recall that in our seismic experiments, we record on a spatially-distributed set of receivers from one shot, and then the pattern is repeated. All these data are recorded digitally and are made to go through a set of procedures called *processing* essentially to convert the raw seismic traces into interpretable seismic images. Although the primary goal of seismic data processing is to obtain realistic images of the subsurface, the initial processing is aimed at noise reduction, employing processes such as trace editing, filtering, and removal of predictable noise in seismic data.

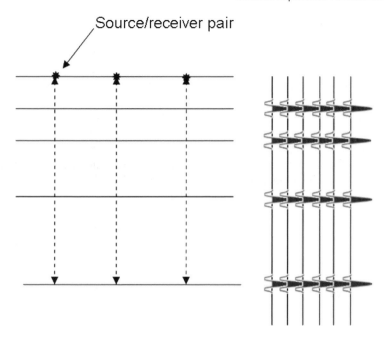

Fig. 1.8—The left panel shows a layered-Earth model and normal incident (coincident source-receiver) ray paths. Plotting the reflection response from several surface locations generates the picture in the right panel. For a complex geologic environment, such a plot is expected to depict subsurface structural variations.

The basic processing steps leading to obtaining seismic images (e.g., Yilmaz 2001) are as follows:

- Geometry.
- Common midpoint (CMP) common depth point (CDP) sorting.
- Velocity analysis.
- Normal moveout (NMO) correction—flatten reflection events.
- Stacking.
- Migration.

1.6.1 Geometry and CMP Sorting. The first step in seismic data processing is to ascertain the true survey geometry involving registration of shot and receiver coordinates and time delays, if any. A "shot gather" contains traces from all the receivers corresponding to a single shot. A very common processing step is to mix traces from all the shots into what is known as a *CMP gather* (**Fig. 1.9**), in which all the traces have a common midpoint between the shot and the receiver. The principal advantage to such a gather is that for moderately heterogeneous Earth structures, the Earth below a midpoint can be approximated very closely with a 1D model (a model in which properties vary with depth only). A CMP sorted gather is shown in Fig. 1.9b. The reflection events in a gather are clearly visible from their near-hyperbolic trajectories.

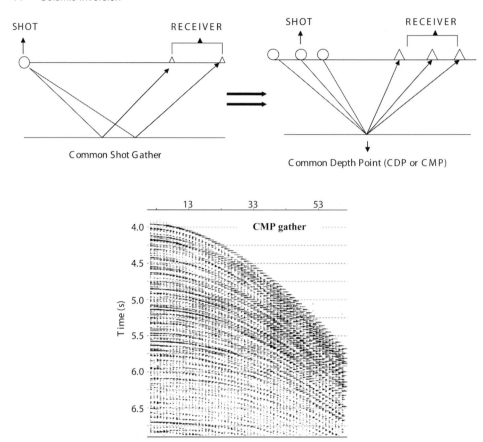

Fig. 1.9—The upper panel shows shot and CMP geometries. The lower panel shows a CMP gather.

1.6.2 Velocity Analysis, Normal Moveout, and Stack. The next step in seismic data processing is to employ certain mathematical operations to CMP gathers to approximate a zero-offset section. Note that in a CMP gather, the arrival time of each reflection event increases with increasing source-receiver offset (a phenomenon called normal moveout—NMO), which can only be predicted if the velocities of the subsurface rock layers are known. For a single interface, we can employ a simple geometric analysis to derive a simple expression for travel time,

$$t^2 = t_o^2 + \left(\frac{x_h}{V}\right)^2, \quad \dots\dots\dots\dots\dots\dots\dots\dots\dots\dots\dots\dots\dots\dots\dots\dots\dots (1.8)$$

where t_o is the two-way normal reflection time, x_h is the offset and V is the velocity of the layer. For a single interface, it is trivial to predict the moveout for a given velocity, and we can try several different velocities until we are able to match the moveout in the data. For multiple layers, this is not a trivial task; it requires ray tracing through the stack of layers.

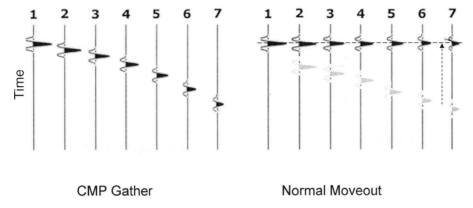

CMP Gather   **Normal Moveout**

Fig. 1.10—Principle of normal moveout correction: time samples from every trace correspon-ding to a nonzero source-receiver offset is aligned with the two-way time of the zero-offset trace for a reflection.

An elegant approximate expression commonly employed to achieve moveout analysis for a stack of layers is given by

$$t^2 = t_o^2 + \left(\frac{x_h}{V_{rms}(t_o^2)} \right)^2, \qquad \dots\dots\dots\dots\dots\dots\dots\dots\dots\dots\dots\dots\dots (1.9)$$

where t_o = two-way zero-offset time, x_h = offset, and V_{rms} = root mean square velocity. This is an approximate equation that is valid for small offsets only; it makes use of V_{rms} such that the inhomogeneous stack of layers is replaced by an equivalent homogeneous layer with the velocity V_{rms}. This entails a very fast computation of travel time in that the complicated ray-tracing procedure is completely avoided here.

For a given reflection event corresponding to a fixed two-way time, the above expression is used to predict travel time (or moveout) for a given offset. The amplitude corresponding to that time sample from the corresponding trace is moved to the time sample correspon-ding to the two-way time—a procedure known as NMO. **Fig. 1.10** shows a schematic rep-resentation of moveout procedure for a single layer. Note that only for a certain velocity will the reflection event become horizontal or moveout-corrected. The procedure for esti-mating velocity is known as NMO analysis. For a stack of layers, the procedure is a gener-alization of the scenario represented in Fig. 1.10. For a given interface (given by its two-way time) and a given set of trial *rms* velocities, the data are moveout-corrected, and the goodness is measured by the coherency of their alignment. A composite coherency plot is generated, from which the point of maximum coherency is picked interactively. This results in an *rms* velocity model and an NMO-corrected gather. This process is repeated for many CMP gathers along a seismic line.

The NMO-corrected gathers are summed or stacked horizontally—the display of all the NMO-corrected and stacked traces is called a *stack section,* and it approximates data that would have been recorded in a zero-offset experiment. In other words, a stack section ap-proximates a zero-offset section. In addition, it enhances the signal-to-noise ratio and re-duces the data volume significantly. An example of velocity analysis and stacking on a seis-mic line is shown in **Fig. 1.11.**

NMO and Stack Process

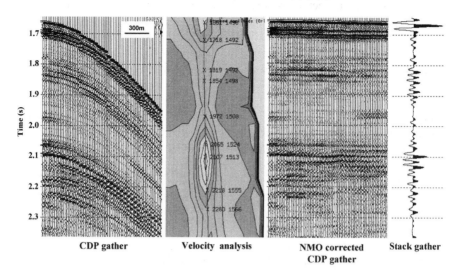

CDP gather   Velocity analysis   NMO corrected   Stack gather
                                    CDP gather

Data: HR6 (Hydrate Ridge), CDP 1739

Fig. 1.11a—Demonstration of velocity analysis and stack on a field CMP gather (left panel). The middle panel shows the semblance plot from which velocities are picked. Proper choice of velocity flattens the reflection events, causing coherence stacking (right panel).

2D Seismic Section (HR 6)

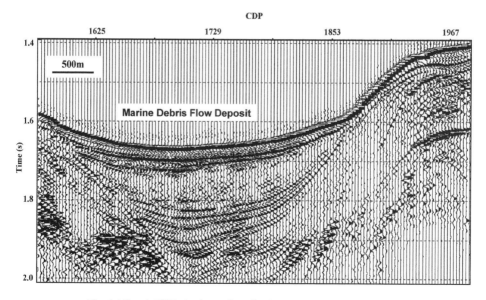

Fig. 1.11b—A CMP stack section displays structural complexity.

1.6.3 Migration. Seismic time sections generated by NMO and stacking are simply rough approximations of geologic sections and are, in general, difficult to interpret in complex areas. In such sections, irrespective of where in space the reflection actually occurs, each event is plotted directly beneath the source-receiver midpoint. Therefore, the location of a dipping reflector is shifted downdip from its true location, concave surfaces may appear to be convex, and spurious faults may appear in the stacked section. In addition, diffractions from curved interfaces and discontinuities are superimposed on targets, making the interpretation process extremely difficult.

Migration is a process of mapping reflection energy back to its true subsurface location. It is a process that corrects the distortions of the geologic structure inherent in the seismic section. In other words, migration attempts to obtain realistic images as far as possible. The mathematics of seismic migration is based on the wave propagation principles discussed in the second chapter. However, a complete description of this vast subject is beyond the scope of this text. We refer interested readers to the text by Yilmaz (2001) for a comprehensive description of the seismic processing algorithms. We simply note here that migration can be applied either to *post-stack* or *prestack* data. For weak lateral variations, a post-stack migration is sufficient to move the reflectors to their true subsurface locations and remove the diffracted energy. However, for complex geologic environments, a prestack migration is employed that makes use of data before they are stacked, and NMO is carried out as a part of the migration process.

Chapter 2

Seismic Wave Propagation

In a seismic experiment, the application of sudden external forces (called *stress*) on solid materials such as the rock layers causes temporal deformation of rocks (called *strain*). Elementary theory of elasticity is generally used to study the propagation of such disturbances through the Earth. A perfectly elastic solid returns to its original shape once the force is removed. In the context of exploration seismology, the Earth is often considered to be perfectly elastic, because the stresses generated by seismic exploration activities are too small to deform subsurface rocks permanently. An explosion source used in seismic exploration injects an impulsive stress over a finite area in the immediate neighborhood of the shot hole, which generates strain in the immediately adjacent subvolume. The strained subvolume then transfers stress to adjacent interior areas within the solid, which generates strains in the surrounding subvolumes. In this manner, an impulsive stress propagates through a solid; the propagation is called an *elastic wave*. Waves consist of a disturbance in materials (media), which carry energy and propagate. However, the material through which the wave propagates does not, in general, move with the wave. The movement of the material is generally confined to small motions—called *particle motions*—of the material as the wave passes. After the wave has passed, the material returns to its original shape and location prior to the passage of the wave. A source of energy such as an explosion creates the initial disturbance (or continuously generates a disturbance as in the case of a Vibroseis source), and the resulting waves propagate out from the disturbance. Because there is finite energy in a confined or short-duration disturbance, the waves generated by such a source will spread out during propagation and become weaker (attenuate) with distance away from the source or with time after the initial source event, and thus, will eventually die out. In the subsurface, we observe various characteristics of such seismic waves. Relationships between stress, strain, and velocity of propagation lead to the wave equations that are fundamental to understanding the behavior of Earth materials and to the analysis of seismograms for resource estimation.

2.1 Stress, Strain, and Hooke's Law

Stress is a measure of force; it is defined as the *force per unit area*. Considering a small cube within the deformed mass we can identify nine components of stress, each of which is given uniquely by the direction of the force and the plane on which the force is acting (**Fig. 2.1a**). When the force acts in a direction perpendicular to a face of the cube shown in Fig. 2.1a, it is called a *normal stress*. Similarly, when the force is tangential or parallel to a face of the cube, it is termed *shear stress*. The stress denoted by the symbol τ is a second-order tensor* with nine scalar components, each of which is characterized by two directions—the direction of the force and the orientation of the plane on which the force is acting. However, because of the equilibrium of forces within a small volume, the off-diagonal elements are symmetric, resulting in six independent components of stress.

The strain is a measure of deformation and is defined as the fractional change in a dimension of a body that results from the application of stress. It is denoted by the symbol ε and is a second-order tensor with six independent components, as shown in Fig. 2.1b. The components of the strain tensor are given in terms of a particle displacement vector $\mathbf{u}$ by the equation

$$\varepsilon = \frac{1}{2}[\nabla\mathbf{u} + (\nabla\mathbf{u})^T], \quad \dots\dots\dots\dots\dots\dots\dots\dots\dots\dots\dots\dots (2.1)$$

or, in indicial notation in a Cartesian coordinate system, as

$$\varepsilon_{kl} = \frac{1}{2}(u_{k,l} + u_{l,k}).$$

Hooke's law relates the strains to the stress; it states that at sufficiently small strains, the strain is directly proportional to the stress producing it. Mathematically, the relationship is expressed as

$$\tau = \mathbf{C}\varepsilon, \quad \dots\dots\dots\dots\dots\dots\dots\dots\dots\dots\dots\dots\dots\dots\dots (2.2)$$

or, in indicial notation,

$$\tau_{ij} = C_{ijkl}\varepsilon_{kl},$$

where $\mathbf{C}$ is a constant characteristic of the medium and the subscripts i, j, k, and l can take values of 1, 2, and 3. In general, $\mathbf{C}$ is a fourth-order tensor (called *elastic coefficient tensor*) consisting of 81 elastic constants. However, because of the symmetry relations of stress, strain, and strain energy, there can at most be 21 nonzero independent elastic constants in a general anisotropic medium (Aki and Richards 2002). The number of elastic coefficients reduces as the number of symmetry relations increases in the medium. An isotropic medium, in which wave velocities are independent of the directions of propagation, is characterized by only two independent elastic constants, λ and μ, known as the *Lamé parameters*. In such a medium, the elastic coefficient tensor is given by

$$C_{ijkl} = \lambda\delta_{ij}\delta_{kl} + \mu(\delta_{ik}\delta_{jl} + \delta_{il}\delta_{jk}), \quad \dots\dots\dots\dots\dots\dots\dots\dots\dots (2.3)$$

where

---

* A tensor of order n is a scalar valued function of n vector variables.

Stress Tensor

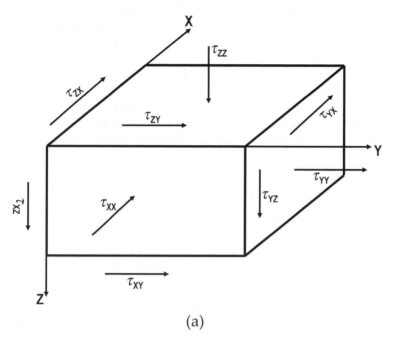

(a)

Strain (deformation) Tensor

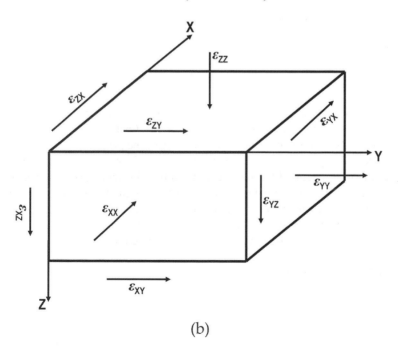

(b)

Fig. 2.1—Different components of (a) stress and (b) strain tensors.

$$\delta_{ij} = 1 \quad \text{for } i = j$$
$$= 0 \quad \text{for } i \neq j,$$

and a summation over repeated indices is implied in Eq. 2.3. Combining Eq. 2.1, 2.2, and 2.3, we obtain the following form of Hooke's law, valid in isotropic media:

$$\boldsymbol{\tau} = \lambda(\nabla \cdot \mathbf{u})\mathbf{I} + \mu[\nabla\mathbf{u} + (\nabla\mathbf{u})^T], \qquad \dots\dots\dots\dots\dots\dots\dots\dots\dots\dots (2.4)$$

where $\mathbf{I}$ is the second-rank identity tensor.

We note here that much seismic data analysis is done assuming an isotropic Earth model and, unless otherwise stated, we will restrict all of our discussions within this text to isotropic media. Many modern seismic data analysis methods do indeed take into account the effects of anisotropy (e.g., Thomsen 1986) but the discussion of such a vast and growing discipline is beyond the scope of this text.

2.2 Equation of Motion

In the definitions of stress and strain tensors, we consider a cubic element in the medium to be bounded by faces that are parallel to the three coordinate planes in a Cartesian axis system (**Fig. 2.2**). The stresses acting on each face are balanced such that the volume is in equilibrium. We must also include inertial terms and body forces in calculating the balance of forces. If $\mathbf{f}$ is the body force per unit volume and ρ is the mass density, the simple application of Newton's second law of motion gives the following equation:

$$\rho \frac{\partial^2 \mathbf{u}}{\partial t^2} = \nabla \cdot \boldsymbol{\tau} + \mathbf{f}, \qquad \dots\dots\dots\dots\dots\dots\dots\dots\dots\dots\dots (2.5)$$

or, in indicial notation,

$$\rho \frac{\partial^2 u_i}{\partial t^2} = \frac{\partial \tau_{ij}}{\partial x_j} + f_i,$$

where $\mathbf{u}(x, y, z, t)$ is the displacement vector, which is a function of the spatial coordinates and time. The inertial terms on the left hand side of the above equation relate density-weighted accelerations to body forces and stress gradients. The above equation is known as the *equation of motion*.

2.3 The Wave Equation

Hooke's law (Eq. 2.4) and the equation of motion (Eq. 2.5) are the two most fundamental equations in seismology. Together they govern the propagation of waves in elastic media, in that the components of stress and displacement as a function of space and time can be computed using these two equations when the description of the body force and the elastic coefficients of the medium are given. A wave equation valid for inhomogeneous elastic media is generally derived by combining these two equations:

$$\rho \frac{\partial^2 \mathbf{u}}{\partial t^2} = \nabla \cdot \{\lambda(\nabla \cdot \mathbf{u})\mathbf{I} + \mu[\nabla\mathbf{u} + (\nabla\mathbf{u})^T]\} + \mathbf{f}$$

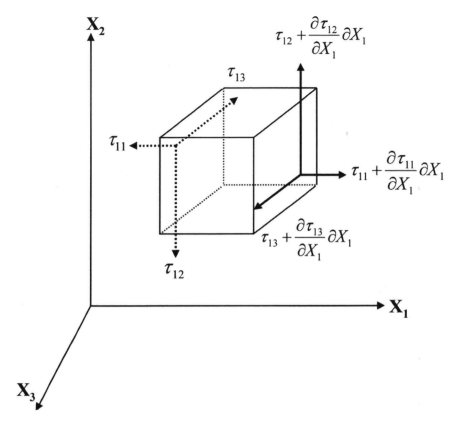

Fig. 2.2—Forces acting on different faces of a cube: balance of forces in each face results in the equation of motion.

or

$$\rho \frac{\partial^2 \mathbf{u}}{\partial t^2} = (\lambda + 2\mu)\nabla(\nabla \cdot \mathbf{u}) - \mu(\nabla \times \nabla \times \mathbf{u}) + \nabla\lambda(\nabla \cdot \mathbf{u})$$

$$+ \nabla\mu \cdot [\nabla\mathbf{u} + (\nabla\mathbf{u})^T] + \mathbf{f}. \qquad \dots\dots\dots\dots\dots\dots\dots (2.6)$$

For a homogeneous elastic medium, the above equation reduces to the following simple form:

$$\rho \frac{\partial^2 \mathbf{u}}{\partial t^2} = (\lambda + 2\mu)\nabla(\nabla \cdot \mathbf{u}) - \mu(\nabla \times \nabla \times \mathbf{u}) + \mathbf{f}. \qquad \dots\dots\dots\dots\dots (2.7)$$

Eqs. 2.6 and 2.7 are the most commonly used forms of the wave equation in seismic data analysis. Yet another, simpler form can be derived by assuming that the medium is acoustic [i.e., by setting the shear modulus to zero ($\mu = 0$)]. An acoustic wave equation is commonly used in seismic data processing involving migration.

2.4 Wave Types

In order to identify common wave types, we make use of the well-known *Helmholtz theorem* that states that any vector field **u** which vanishes as |**r**|→∞ (where **r** is the distance vector) can be represented as a sum of a vector potential **ψ** and a scalar potential ϕ such that

$$\mathbf{u} = \nabla\phi + \nabla \times \boldsymbol{\psi}, \qquad \dots\dots\dots\dots\dots\dots\dots\dots\dots\dots (2.8)$$

with

$$\nabla \bullet \boldsymbol{\psi} = 0.$$

We make use of the potential representation of the displacement field in the wave equation for homogeneous media and set **f** = 0 in Eq. 2.7 to obtain the following equation:

$$\nabla\left[(\lambda + 2\mu)\nabla^2\phi - \rho\frac{\partial^2\phi}{\partial t^2}\right] + \nabla \times \left[\mu\nabla^2\boldsymbol{\psi} - \rho\frac{\partial^2\boldsymbol{\psi}}{\partial t^2}\right] = 0, \qquad \dots\dots\dots\dots (2.9)$$

which gives

$$\nabla^2\phi - \frac{\rho}{(\lambda + 2\mu)}\frac{\partial^2\phi}{\partial t^2} = 0, \qquad \dots\dots\dots\dots\dots\dots\dots\dots\dots (2.10)$$

and

$$\nabla^2\boldsymbol{\psi} - \frac{\rho}{\mu}\frac{\partial^2\boldsymbol{\psi}}{\partial t^2} = 0.$$

Thus, we obtain a scalar wave equation for potential ϕ and a vector wave equation for potential **ψ** with coefficients

$$\frac{1}{\alpha^2} \text{ and } \frac{1}{\beta^2},$$

respectively, such that

$$\nabla^2\phi - \frac{1}{\alpha^2}\frac{\partial^2\phi}{\partial t^2} = 0, \qquad \dots\dots\dots\dots\dots\dots\dots\dots\dots (2.11)$$

and

$$\nabla^2\boldsymbol{\psi} - \frac{1}{\beta^2}\frac{\partial^2\boldsymbol{\psi}}{\partial t^2} = 0,$$

where

$$\alpha = \sqrt{\frac{\lambda + 2\mu}{\rho}}, \quad \dots\dots\dots\dots\dots\dots\dots\dots\dots\dots\dots\dots (2.12)$$

and

$$\beta = \sqrt{\frac{\mu}{\rho}}.$$

As is evident from the definition of the potentials, the displacement field comprises two fundamental wave types corresponding to the two potentials. For example, the scalar potential ϕ represents compressional motion and volumetric changes, while the vector potential ψ represents shearing motion with no volume change (*zero divergence*). They also propagate with two distinct velocities given by the coefficients α and β. These two wave-types are called Compressional (or Primary) or *P-waves*, and Shear or *S-waves*, which are described below.

The P-waves represent compressional motion, in that the particles move along the direction of propagation, and the S-waves are shear waves, in which the particle motion is always orthogonal to the direction of propagation. Note that such pure P and S modes are observable in a homogeneous medium. It can also be shown that even in heterogeneous media, P and S motions are distinct at high frequencies. For all rock types in isotropic media, the P-wave velocity is greater than the shear-wave velocity and, therefore, the P-waves arrive earlier than the corresponding S-waves. Even though other wave-types— such as Rayleigh and Love waves (*surface waves*)—are observable in an elastic Earth, most seismic data analysis for exploration applications can be carried out using only the two body waves, the P- and S-waves. It is also worthwhile to note that two types of shear motions are observable, one that is polarized perpendicular to the direction of propagation in the plane of propagation, called the *SV-wave*, and the other, polarized normally to the direction of propagation and the plane of propagation, called the *SH-wave*. **Fig. 2.3** describes the P and S motions. As the names imply, the shear modulus, μ, is proportional to the shear strength of the medium, while the Lamé parameter, λ, describes the incompressibility of the medium. The elastic constants have real values and are never negative, so a comparison of the expressions for α and β (Eq. 2.12) supports the observation that P-waves always travel faster than S-waves. Also, the expression for Eq. 2.12 implies that only solid bodies with shear strength can support shear-wave propagation. Fluids such as air and water do not support the propagation of S-waves (because $\mu = 0$).

2.5 Spherical Waves and Plane Waves

Let us consider a scalar wave equation with a scalar body force term that is localized in space and time, given by

$$\frac{\partial^2 \phi}{\partial t^2} = \alpha^2 \nabla^2 \phi + \delta(\mathbf{x})\delta(t), \quad \dots\dots\dots\dots\dots\dots\dots\dots\dots\dots\dots\dots (2.13)$$

where δ represents a dirac-delta function and $\mathbf{x}$ is the position vector. Aki and Richards (2002) and others have shown that the solution of this scalar wave equation is given by

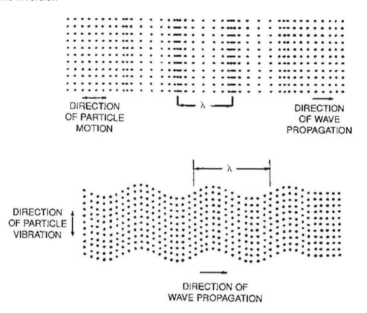

Fig. 2.3—Particle motion direction with respect to the direction of wave propagation for P-waves (upper panel) and S-waves (lower panel) (from Lay and Wallace, 1995; reprinted with permission from Academic Press).

$$\phi(\mathbf{x},t) = \frac{1}{4\pi\alpha^2} \frac{\delta\left(t - \frac{|\mathbf{x}|}{\alpha}\right)}{|\mathbf{x}|}. \qquad \dots\dots\dots\dots\dots\dots\dots\dots\dots\dots\dots(2.14)$$

Note the spherical symmetry of the solution of this scalar wave equation where $|\mathbf{x}|$ measures the distance of a point in space from the source location. The solution possesses the following properties:

- The amplitude of ϕ decays with the reciprocal of the distance away from the source.
- The waves arrive at all points at constant distance away from the source at the same time.

In other words, in homogeneous media, the waves propagate away from the source as spherical wavefronts, the amplitude of which decays with increasing distance in a phenomenon called *geometrical spreading*.

Fig. 2.4 shows cross sections of spherical wavefronts as circles; every point on a circle receives energy at exactly the same time. Note that far away from the source, the radius of the sphere becomes so large that we can approximate the wavefront as a plane wave. Plane waves play a very important role in the theoretical development of wave propagation equations that are used routinely in seismic data analysis. It is intuitively obvious that we can draw a planar wavefront at each point on a spherical wavefront that is tangent to the spherical wavefront. Each tangent plane is uniquely described by the angle it makes with the ver-

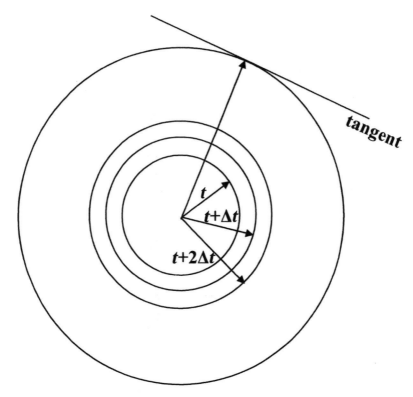

Fig. 2.4—Vertical cross-section of a spherical waveform in a homogeneous isotropic medium: at a large distance away from the source, the radius becomes so large that the wavefronts appear planar. Each tangent plane is uniquely described by the angle it makes with the vertical or horizontal at the source location.

tical (or horizontal) at the source. A time series can be constructed by a summation of individual frequency components using a Fourier series. Similarly, a spherical wavefront can be constructed by a summation of planar wavefronts.

At this stage, we recall the definition of a temporal Fourier transform given by the pair

$$\tilde{f}(\omega) = \int_{-\infty}^{\infty} f(t)\exp(i\omega t)dt, \quad \dots\dots\dots\dots\dots\dots\dots\dots\dots (2.15)$$

and

$$f(t) = \frac{1}{2\pi}\int_{-\infty}^{\infty} \tilde{f}(\omega)\exp(-i\omega t)d\omega,$$

where ω is the temporal frequency. Similarly, a spatial Fourier transform is given by

$$\tilde{f}(k_x) = \int_{-\infty}^{\infty} f(x)\exp(-ik_x x)dx, \qquad \ldots\ldots\ldots\ldots\ldots\ldots\ldots\ldots\ldots (2.16)$$

where x is the space coordinate and k_x is the x-component of the wave number. Thus, a propagating variable of interest $\phi(x,y,z,t)$ can be expressed in terms of its spatial and temporal frequency components by the following multidimensional integral:

$$\phi(x,y,z,t) = \frac{1}{8\pi^2} \int_{-\infty}^{\infty} dk_x \int_{-\infty}^{\infty} dk_y \int_{-\infty}^{\infty} d\omega \phi(k_x,k_y,z,\omega)\exp[i(k_x x + k_y y - \omega t)], \qquad \ldots\ldots 2.17)$$

Thus, our spherical wave solution (Eq. 2.14) can be expressed as

$$\frac{1}{|\mathbf{x}|}\exp\left[i\omega\left(\frac{|\mathbf{x}|}{\alpha} - t\right)\right] = \frac{\exp(i\omega t)}{2\pi^2} \int_{-\infty}^{\infty}\int_{-\infty}^{\infty}\int_{-\infty}^{\infty} \frac{\exp(i\mathbf{k} \bullet \mathbf{x})}{k^2 - \frac{\omega^2}{\alpha^2}} dk_x dk_y dk_z, \qquad \ldots\ldots\ldots (2.18)$$

or

$$\frac{1}{|\mathbf{x}|}\exp\left(i\omega\left(\frac{|\mathbf{x}|}{\alpha} - t\right)\right) = \frac{\exp(-i\omega t)}{2\pi^2} \int_{-\infty}^{\infty}\int_{-\infty}^{\infty}\int_{-\infty}^{\infty} A\exp(ik_x x + ik_y y + ik_z z)dk_x dk_y dk_z, \quad \ldots (2.19)$$

where the vector wave number $\mathbf{k}$ is given by

$$\mathbf{k} = k_x \hat{x} + k_y \hat{y} + k_z \hat{z}, \qquad \ldots\ldots\ldots\ldots\ldots\ldots\ldots\ldots\ldots\ldots\ldots (2.20)$$

and

$$|\mathbf{k}| = \sqrt{k_x^2 + k_y^2 + k_z^2} = \frac{\omega}{\alpha}. \qquad \ldots\ldots\ldots\ldots\ldots\ldots\ldots\ldots$$

Note that the function containing an exponential (called the *phase term*) in the integrand $\exp[i(\mathbf{k} \bullet \mathbf{x} - \omega t)]$ in Eq. 2.19 represents a plane wave free to propagate in any direction in the continuum. This is because for a given frequency ω and velocity α, the requirement that

$$|\mathbf{k}| = \sqrt{k_x^2 + k_y^2 + k_z^2} = \frac{\omega}{\alpha}$$

is a constant defines a planar surface in Cartesian space with a normal vector

$$\mathbf{k} = |\mathbf{k}|\hat{k} = \frac{\omega}{\alpha}\hat{k},$$

called a *wave number vector* (**Fig. 2.5**). At this stage, we define a slowness vector as the reciprocal of velocity; i.e.,

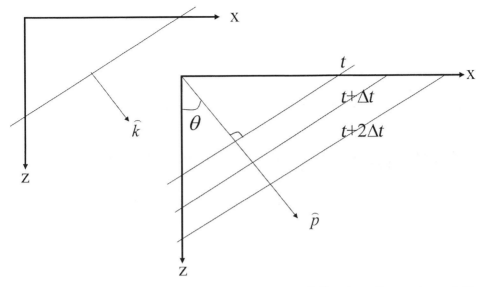

Fig. 2.5—A plane wave is defined by a wave number vector k̂. The planar time contours at different time steps are shown in the lower panel.

$$\mathbf{p} = \frac{1}{\alpha}\hat{p} \qquad \dots\dots\dots\dots\dots\dots\dots\dots\dots\dots\dots\dots\dots (2.21)$$

such that the phase term is given by

$$\exp[i\omega(\mathbf{p} \bullet \mathbf{x} - t)], \text{ where } \mathbf{k} = \omega\mathbf{p} \text{ and } |\mathbf{p}|\sqrt{p_x^2 + p_y^2 + p_z^2}.$$

In two dimensions, only the horizontal component p_x exists and is denoted by p and the vertical component of the slowness is denoted by q; they are given by

$$p = \frac{\sin\theta}{\alpha}, q = \frac{\cos\theta}{\alpha}, \text{ and } \frac{1}{\alpha^2} = p^2 + q^2, \qquad \dots\dots\dots\dots\dots\dots\dots (2.22)$$

where θ is the angle measured from the vertical (Fig. 2.5).

Note that the exponential function $\exp[i\omega(\mathbf{p} \bullet \mathbf{x} - t)]$ represents an oscillatory motion. At a constant frequency, constant time contours are planar in three dimensions (hence the name *plane waves*) and linear in 2D, as shown in Fig. 2.5.

2.6 Wavefronts and Rays

A wavefront is a surface on which all points are in the same phase of vibration (i.e., have equal travel time). In a homogeneous medium, the wavefronts are spherical, and at a large distance away from the source, they appear planar. Spherical waves can also be constructed by a superposition of plane waves propagating at all angles. For a line source in a homogeneous medium, the wavefronts are cylindrical, and such cylindrical waves can be constructed from a superposition of plane waves. Thus, much like the basic role of frequency in time series analysis, plane waves are the basic building blocks upon which wave propagation theories are built. In a homogeneous medium, travel time can be computed simply by multiplying the source-receiver distance (a straight line) with the reciprocal of the velocity. In an inhomogeneous medium, waves can travel along numerous paths. **Fermat's principle** states that the waves travel along a path such that the travel time between two points is the minimum. Thus, wavefronts no longer remain spherical in a heterogeneous medium.

Seismic wave behavior can sometimes be understood intuitively in terms of ray theory. Rigorous development of ray theory is based upon an asymptotic (or high-frequency) solution of the elastic wave equation (e.g., Cerveny 2001). In an isotropic medium, a ray path is always perpendicular to the wavefronts, is a minimum time path, and is a path along which the energy propagates. Rays deviate from straight-line paths in a heterogeneous medium. For example, consider a medium consisting of flat layers with each layer characterized by a unique P-wave velocity (**Fig. 2.6**), and with the layers separated by discontinuities in P-wave velocity. The ray paths through the stack of layers (or in a general heterogeneous medium) are governed by Snell's law, which dictates that the horizontal slowness is preserved. Recall that the horizontal slowness is given by

$$p = \frac{\sin\theta}{\alpha}.$$

Therefore, for the stack of layers shown in Fig. 2.6, we have

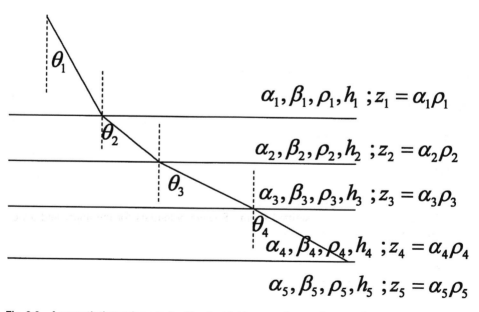

Fig. 2.6—A raypath through a stack of horizontal layers—the angles are given by Snell's law. α_i, β_i, ρ_i, h_i, and z_i are the P-wave velocity, S-wave velocity, density, thickness, and acoustic impedance, respectively, of layer i.

$$p = \frac{\sin \theta_1}{\alpha_1} = \frac{\sin \theta_2}{\alpha_2} = \frac{\sin \theta_3}{\alpha_3} = \ldots = \text{constant.} \qquad \ldots\ldots\ldots\ldots\ldots\ldots\ldots (2.23)$$

2.7 Plane-Wave Reflection at a Boundary

As a first step to studying wave propagation in heterogeneous media, we consider the interaction of a plane wave at an interface across which elastic properties such as the P-wave velocity, S-wave velocity, and density changes are discontinuous. Most seismic analysis principles (such as CMP and NMO, described in Chapter 1), to a first order approximation, assume the rock layers to be laterally homogeneous, separated by planar interfaces (Fig. 2.6). Understanding the interaction of plane waves at material discontinuities is fundamental to developing complex and realistic wave propagation models.

The principle of interaction of plane waves at a planar interface between two rock layers with contrasting elastic properties is governed by the boundary conditions described below.

- At a welded contact, all components of displacement and the normal components of stress are continuous.
- At a solid/fluid interface, the normal components of stress and normal components of displacement are continuous.

It can be shown that an incident P-plane wave at an elastic interface gets reflected both as a P-wave and an S-wave and that there exist both a transmitted P-wave and a transmitted S-wave in the lower layer (Aki and Richards 2002). As shown in **Fig. 2.7**, the reflection and transmission of P- and S-waves satisfy Snell's law such that

$$p = \frac{\sin i_1}{\alpha_1} = \frac{\sin i_2}{\alpha_2} = \frac{\sin j_1}{\beta_1} = \frac{\sin j_2}{\beta_2}. \qquad \ldots\ldots\ldots\ldots\ldots\ldots\ldots\ldots (2.24)$$

The above equations clearly demonstrate that the angle made by the incident P-wave ray is the same as the angle made by the reflected P-wave ray in the upper layer.

In addition to changes in the direction of the ray path, the amplitudes of the reflected and transmitted P- and S-waves are also expected to exhibit changes, because of partitioning of energy at the interface. The *reflection and transmission coefficients* of PP and PS waves, denoted by *Rpp, Rps, Tpp,* and *Tps*, measure the amplitudes of reflected and transmitted P- and S-waves because of an incident plane wave of unit amplitude. The recipe for deriving expressions for reflection and transmission coefficients is given below.

- Define a unit amplitude incident plane wave for P-wave potential.
- Define reflected and transmitted P- and S-wave potentials for the upper and lower layer with amplitudes *Rpp, Rps, Tpp,* and *Tps,* as appropriate.
- Express the displacements in the upper and lower layers using Helmholtz's theorem.
- Express stresses in terms of displacements using Hooke's law.
- Apply the continuity conditions for displacements and normal stress.

The above procedure results in a linear system of equations that can be solved analytically for the reflection and transmission coefficients. The expressions for *Rpp* and *Rps* (Aki and Richards 2002) are given by

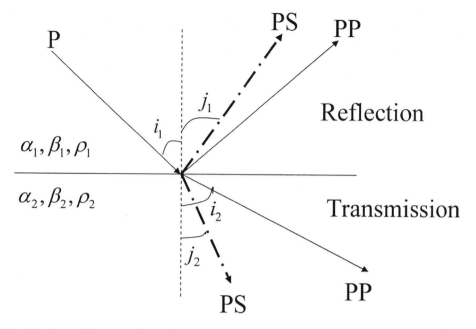

Fig. 2.7—P- and S-wave reflections from and transmission across an interface of material discontinuity.

$$Rpp = \frac{\left[\left(b\dfrac{\cos i_1}{\alpha_1} - c\dfrac{\cos i_2}{\alpha_2}\right)F - \left(a + d\dfrac{\cos i_1}{\alpha_1}\right)\dfrac{\cos j_2}{\beta_2}Hp^2\right]}{D},$$

$$Rps = -\frac{2\dfrac{\cos i_1}{\alpha_1}\left(ab + cd\dfrac{\cos i_2}{\alpha_2}\dfrac{\cos j_2}{\beta_2}\right)p\alpha_1}{\beta_1 D}, \quad\ldots\ldots\ldots\ldots\ldots\ldots (2.25)$$

where the constants a, b, c, d, F, H, and D are all expressed in terms of the elastic properties of the two layers and the angles of incidence, reflection, and transmission of P-waves. Thus we observe that even though the computation of such reflection coefficients is an easy task on a modern computer, the behavior of these complicated functions is very difficult to predict analytically. From the numerical evaluation of Eq. 2.25, we learn that Rpp and Rps are, in general, complex numbers, the values of which depend on the angles of incidence of the plane waves. At normal incidence ($\theta = 0$; $p = 0$), Rpp reduces to the following simple form:

$$Rpp = \frac{\rho_2\alpha_2 - \rho_1\alpha_1}{\rho_2\alpha_2 + \rho_1\alpha_1}. \quad\ldots\ldots\ldots\ldots\ldots\ldots\ldots\ldots\ldots\ldots\ldots\ldots\ldots\ldots (2.26)$$

The product of compressional wave velocity and density ($\rho\alpha$) is called *acoustic imped-ance* (AI), and it is obvious from Eq. 2.26 that the contrast in AI across an interface determines the amplitude of reflected plane P-waves at normal incidence. Thus, a post-stack section that emulates normal-incidence seismograms can be modeled with Eq. 2.26.

A simpler form of reflection coefficient that is derived from linearization of Eq. 2.25, called a *linearized reflection coefficient*, is often used in amplitude analysis, and it also displays the behavior of the reflection coefficients clearly. Aki and Richards (2002) derived the linearized reflection coefficients, assuming that the contrasts in elastic properties of the media are small; i.e., $\Delta\rho = \rho_2 - \rho_1$, $\Delta\alpha = \alpha_2 - \alpha_1$, and $\Delta\beta = \beta_2 - \beta_1$, and the ratios

$$\frac{\Delta\rho}{\rho}, \frac{\Delta\alpha}{\alpha}, \frac{\Delta\beta}{\beta}$$

are much less than one. The α, β, and ρ are the average values of P-wave velocity, S-wave velocity, and density. Under these assumptions, we obtain the following form of the equation:

$$Rpp(p) \cong \frac{1}{2}(1 - 4\beta^2 p^2)\frac{\Delta\rho}{\rho} + \frac{1}{2\cos^2 i}\frac{\Delta\alpha}{\alpha} - 4\beta^2 p^2 \frac{\Delta\beta}{\beta}, \quad \dots\dots\dots\dots (2.27)$$

in which the contributions of the contrasts in the three parameters can be easily identified. Shuey (1985) derived a similar form of the linearized reflection coefficient given by

$$R_{pp}(i) = R_0 + \left(A_0 R_0 + \frac{\Delta\sigma}{(1 - \sigma)^2}\right)\sin^2 i + \frac{1}{2}\frac{\Delta\alpha}{\alpha}(\tan^2 i - \sin^2 i), \quad \dots\dots (2.28)$$

where

$$A_0 = B_0 - 2(1 + B_0)\left(\frac{1 - 2\sigma}{1 - \sigma}\right)$$

$$B_0 = \frac{\Delta\alpha/\alpha}{\left(\frac{\Delta\alpha}{\alpha} + \frac{\Delta\rho}{\rho}\right)},$$

R_0 is the normal-incidence reflection coefficient, and σ is Poisson's ratio. Eq. 2.28 is sometimes written as

$$R_{pp}(i) = A + B\sin^2 i + C\tan^2 i. \quad \dots\dots\dots\dots\dots\dots\dots\dots (2.29)$$

It is clear from Eq. 2.29 that the effect of shear waves on reflection coefficients is more pronounced at large angles of incidence. At normal incidence, only the A term determines the reflection coefficient; at small angles, the first two terms in Eq. 2.29 affect the reflection coefficient, and at moderate angles, all three terms need to be used. A comparison of two-term reflection coefficients with those using exact equations for three models of shale/sand, shale/gas/sand and shale/salt interfaces (**Table 2.1**) is shown in **Fig. 2.8**. Note that for all three models the two-term linearized reflection coefficients follow the exact reflection coefficients closely within small angles (<20°).

TABLE 2.1—P-WAVE VELOCITY, S-WAVE VELOCITY, AND DENSITY VALUES OF FOUR ROCK TYPES			
Interface	Density (kg/m³)	P-Wave Velocity, α (m/s)	S-Wave Velocity, β (m/s)
Shale	2500	2770	1480
Gas sand	2200	2790	1910
Brine sand	2300	3200	1870
Salt	2200	4270	2470

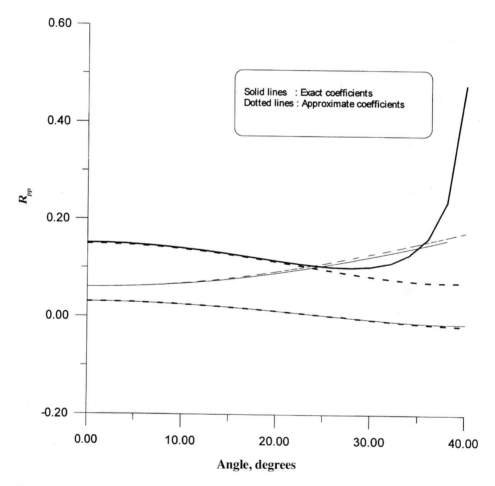

Fig. 2.8—Reflection coefficients as a function of incidence angle of shale/gas sand, shale/brine sand, and shale/salt interfaces (Table 2.1); the solid lines represent exact reflection coefficients, while the dashed lines represent linearized reflection coefficients.

2.8 Seismic Response of a Layered Earth Model

To a first approximation, an Earth model is assumed to consist of flat layers separated by plane boundaries (Fig. 2.6). In a homogeneous medium, waves propagate as spherical wavefronts, and the wave amplitude decays with distance from the source. On the other hand, plane waves do not have geometrical divergence loss in a homogeneous medium, and in an inhomogeneous-layered medium, plane waves undergo reflection and transmission losses affecting their amplitudes as they propagate through the medium. Because a point-source response can be constructed by a superposition of the responses from plane waves at all angles, for layered Earth models we can simply modify the plane-wave responses owing to reflection and transmission through the stack of layers prior to superposition. So far, we have identified two causes of amplitude losses (or modifications), namely, the geometric spreading and reflection and transmission. Before we take up the case of the complete response of the layered Earth model, let us discuss the case of post-stack data.

2.8.1 Post-Stack Modeling. Post-stack data approximate the response of a layered Earth to a normal-incidence plane wave. The normal-incidence reflection coefficient at a planar interface between two layers is given by

$$R_{pp}^1(i = 0) = \frac{Z_2 - Z_1}{Z_2 + Z_1}, \qquad \dots\dots\dots\dots\dots\dots\dots\dots\dots\dots\dots\dots (2.30)$$

where Z_i represents the acoustic impedance of layer i, and R_{pp}^1 represents the P-wave reflection coefficient of the first interface. For a single interface, the amplitude will be given by the reflection coefficient, and the wave arrives at a time given by the two-way reflection time for vertical propagation, also called two-way time (TWT). For a layer i with thickness h_i, the TWT t_i is given by

$$t_i = 2\frac{h_i}{\alpha_i}. \qquad \dots\dots\dots\dots\dots\dots\dots\dots\dots\dots\dots\dots\dots (2.31)$$

The time series or seismogram $S(t)$ corresponding to reflection from one interface is given by

$$S(t) = R_{pp}^1 \delta(t - t_1). \qquad \dots\dots\dots\dots\dots\dots\dots\dots\dots\dots (2.32)$$

Eq. 2.32 simulates the response to a delta function wavelet, but it is impossible to inject that entity into the ground. As discussed in Chapter 1, a band-limited wavelet is often used as an energy source. Thus, we need to modify Eq. 2.32 simply by a process called convolution (Yilmaz 2001), given by

$$S_w(t) = S(t) \otimes W(t) = W(t) \otimes R_{pp}^1 \delta(t - t_1), \qquad \dots\dots\dots\dots\dots\dots\dots\dots (2.33)$$

where $W(t)$ is the wavelet and the symbol $\otimes$ represents a convolution. In the frequency domain, a convolution is equivalent to a multiplication such that Eq. 2.33, after Fourier transformation, reduces to the following form:

$$S_w(\omega) = S(\omega)W(\omega) = W(\omega)R_{pp}^1 \exp(i\omega t_1), \qquad \dots\dots\dots\dots\dots\dots\dots (2.34)$$

where ω is the angular frequency.

Digital convolution of two-time series can be performed using the following steps:

1. Time-reverse one-time series.
2. Apply time shift.
3. Multiply and add.

Let us consider two-time series of four samples each given by

{1.0   0.5   –0.1   0.2} and {4.0   3.0   1.5   2.0}.

We reverse one sequence (e.g., the second sequence) and then, keeping the first sequence fixed, we shift the second sequence in steps and multiply and add as follows:

Shift = 0

		1.0	0.5	–0.1	0.2			
2.0	1.5	3.0	4.0					

= 4.00

Shift = 1

		1.0	0.5	–0.1	0.2		
	2.0	1.5	3.0	4.0			

= 5.00

Shift = 2

		1.0	0.5	–0.1	0.2	
		2.0	1.5	3.0	4.0	

= 2.60

Shift = 3

	1.0	0.5	–0.1	0.2	
	2.0	1.5	3.0	4.0	

= 3.25

Shift = 4

	1.0	0.5	–0.1	0.2	
		2.0	1.5	3.0	4.0

= 1.45

Shift = 5

	1.0	0.5	–0.1	0.2	
			2.0	1.5	3.0	4.0

= 0.10

Shift = 6

	1.0	0.5	–0.1	0.2	
			2.0	1.5	3.0	4.0

= 0.40

The output sequence is the convolution of the two sequences given by {4.00   5.00   2.60   3.25   1.45   0.10   0.40}.

The response of the multilayered Earth can be computed simply by summing the responses of individual layers as computed by Eq. 2.34. Note that there is no conversion from P to S at normal incidence. Assuming that the effect of internal reverberation through individual layers is negligible, post-stack synthetic seismograms for the layered-Earth model is given by

$$S_w(\omega) = W(\omega) \sum_{j=1}^{n} R_{pp}^j \exp(i\omega t_j), \qquad \dots\dots\dots\dots\dots\dots\dots\dots\dots\dots (2.35)$$

where n is the total number of interfaces (total number of layers $= n+1$). Eq. 2.35 needs to be transformed to the time domain by applying a fast Fourier transform (FFT). Note that the spherical spreading is ignored in this case; this assumes that the data to be modeled have been corrected for geometric divergence, a common step in seismic data processing. A more general approach to modeling post-stack data makes use of an exploding reflector model in a finite- difference solution of an acoustic wave equation.

2.8.2 Prestack Modeling. The basic concepts leading to the generation of prestack synthetic seismograms can be understood by following the procedure employed in modeling under the post-stack assumption. However, there are some important differences that need to be addressed. For example,

- We will need to compute responses of all possible plane waves propagating at different angles characterized by their unique ray parameters (Eq. 2.23).
- The reflection coefficients are functions of ray parameters or angles, which can be computed by either Eq. 2.25 or by an approximate form given by Eq. 2.29.
- For each incident P plane wave, conversion to S needs to be considered. Although the amplitude loss through conversion to S is accounted for in the equations for PP reflection coefficients, S-wave arrival paths need to be modeled. That is, even though an explosion source does not generate an S-wave at the source, converted S-waves propagate through the medium and will be recorded at a receiver on the surface of a solid Earth.
- Numerous internal reverberations—and conversions thereof—need to be taken into account.
- Finally, the complete response of all the plane waves must be summed to generate a point-source response that will automatically include the effect of spherical spreading.

As in the post-stack case, the first step to synthesizing seismograms in the prestack domain involves computation of arrival times of different reflection events. For a plane wave characterized by angle θ, the reflection travel time **(Fig. 2.9)** is given by a vertical travel time (also called *delay time*) and a horizontal travel time; i.e.,

t = vertical delay time + horizontal time

$$t = 2 \frac{\cos \theta}{\alpha} h + \frac{\sin \theta}{\alpha} x = 2qh + px = \tau + px, \qquad \dots\dots\dots\dots\dots\dots (2.36)$$

where p and q are the horizontal and vertical slownesses, respectively, and

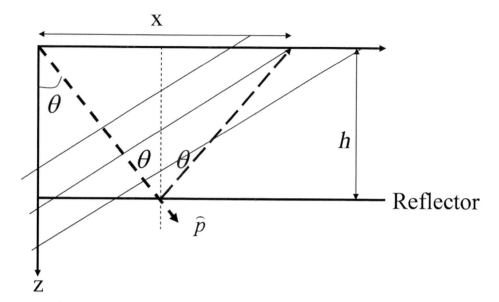

Fig. 2.9—Reflection of a plane-wave incident at an angle θ at an interface of a layer with thickness h.

$$\tau = 2qh$$

is called the *vertical delay time*. In Eq. 2.36, the horizontal time component is added in the transformation from the slowness domain to the offset domain, and therefore, for plane-wave computation, it is sufficient to include the vertical delay time component only. Thus, the plane-wave response for a P-wave reflection from a single interface (ignoring surface interaction) is given by

$$A_{pp}^1 = R_{pp}^1 \delta(t - \tau_{pp}^1).$$

In the frequency domain after multiplication with the spectrum of the wavelet we have

$$S(\omega) = W(\omega)R_{pp}^1 \exp(i\omega\tau_{pp}^1). \qquad \dots\dots\dots\dots\dots\dots\dots\dots (2.37)$$

The evaluation of Eq. 2.37 followed by transformation to offset-time domain results in the desired seismograms. An example of the computation for a single layer over a half-space model (**Fig. 2.10a**) is shown in Fig. 2.10b. Note the variation of amplitude with offset that reflects changes caused by the reflection coefficient and spherical spreading. For elastic computation, we simply need to modify Eq. 2.37 to include a shear-wave path by adding Rps and a corresponding vertical delay-time component that uses shear-wave velocity. The elastic seismograms for vertical geophones are shown in Fig. 2.10c.

The computations shown in Figs. 2.10b and c are far from being complete, because the surface reflections, their multiple reflections and mode conversions for P to S, are not included. For complete synthesis, we need to include all ray paths, including conversions and

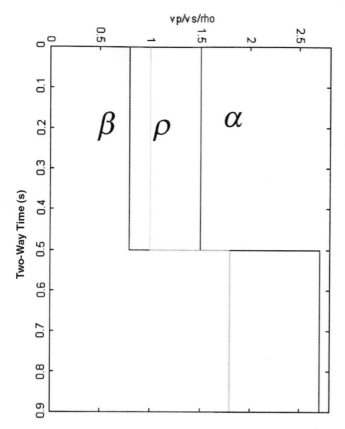

Fig. 2.10a—An Earth model consisting of an elastic layer over an elastic half space; P- and S-wave velocities are in km/s, and the density is in g/cm^3.

internal multiples; there are too many to count. This is, however, achieved through a more rigorous development known as *invariant imbedding* or a *reflectivity* method (Kennett 1983). Complete description of the development of the reflectivity method is beyond the scope of this text. However, a clear understanding of the method is required for data modeling because reflectivity modeling is one of the most popular approaches.

In this method, we start with the equation of motion (Eq. 2.5) in the frequency domain and the stress-strain relation (Eq. 2.4). For a stratified elastic medium, we can transform out the dependence of **u** and $\boldsymbol{\tau}$ on radius and azimuth by expanding them as a series of cylindrical harmonics via the **Fourier-Hankel transformation**. The resulting system of equations has the following form:

$$\partial_z \mathbf{b} = -i\omega \mathbf{A b} + \mathbf{f}, \quad \dots\dots\dots\dots\dots\dots\dots\dots\dots\dots\dots\dots\dots (2.38)$$

where $\mathbf{b} = [u_x\, u_z\, \tau_{xz}\, \tau_{zz}\, u_y\, \tau_{yz}]^T$ is called a *stress-displacement vector,* the elements of which are the scalar components of displacement and tractions. The matrix $\mathbf{A}(\omega, p)$, called the *system matrix,* is a function of elastic coefficients, and **f** is a body force term. For isotropic and transversely isotropic media, the equation (2.38) decouples into two systems, namely a

P-wave Primaries only

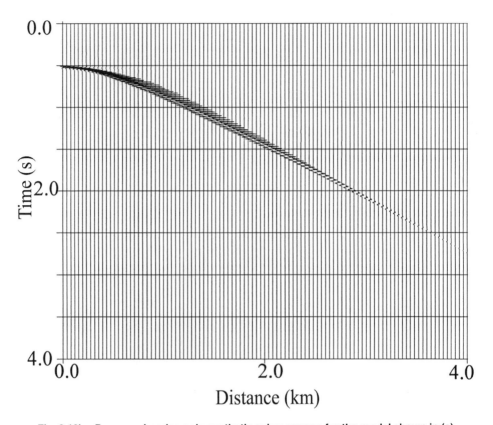

Fig. 2.10b—P-wave primaries only synthetic seismograms for the model shown in (a).

P-SV (4×4) system and an SH (2×2) system. The solution of the system of equations (Eq. 2.38) can be carried out by a propagator matrix method. It is well known that the propagator matrix is generally unstable because of growing exponentials. Stable solutions can be obtained by one of three methods:

- A global matrix approach.
- A compound matrix approach in which we define a new system of ODE in which the elements of the new system matrix are the minors of the original system matrix. The original 4×4 P-SV system maps into a 6×6 system.
- An invariant imbedding or reflection matrix approach (Kennett 1983), popularly known as the reflectivity method.

Of the three methods listed above, the compound matrix and the reflectivity methods have been widely used in seismology. The reflectivity method has been very popular because of its ray-interpretation and easy generalization to azimuthally anisotropic media. In the unconditionally stable reflection-matrix approach (Kennett 1983), the propagation uses the

P- and S- waves (Primaries only)

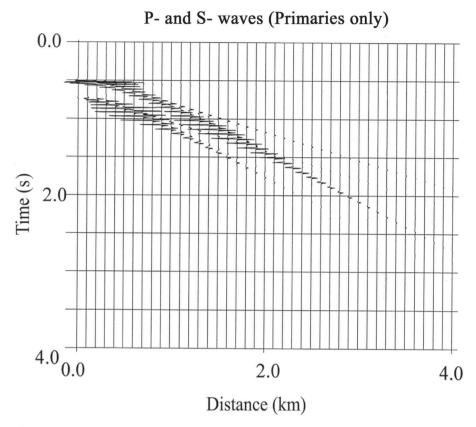

Fig. 2.10c—Primary P- and S-waves only synthetic seismograms for the model shown in (a).

eigenvalues and eigenvectors of the system matrix $\mathbf{A}$. These eigenvalues and eigenvectors are used to define four up going and down going reflection- and transmission-coefficient matrices, $\mathbf{R}_D$, $\mathbf{T}_D$, $\mathbf{R}_U$, and $\mathbf{T}_U$. These are propagated through the stack of layers to obtain a composite reflection matrix that includes the effects of reflection, transmission, mode-conversion, and internal multiples. Kennett (1983, p.127) derived the following iteration equation, which can be used to compute the up going and down going reflection-and-transmission matrices of a zone AC (**Fig. 2.11**) when those of zones AB and BC are known:

$$\mathbf{R}_D^{AC} = \mathbf{R}_D^{AB} + \mathbf{T}_U^{AB}\mathbf{R}_D^{BC}[\mathbf{I} - \mathbf{R}_U^{AB}\mathbf{R}_D^{BC}]^{-1}\mathbf{T}_D^{AB}$$

$$\mathbf{T}_D^{AC} = \mathbf{T}_D^{BC}[\mathbf{I} - \mathbf{R}_U^{AB}\mathbf{R}_D^{BC}]^{-1}\mathbf{T}_D^{AB}$$

$$\mathbf{R}_U^{AC} = \mathbf{R}_U^{BC} + \mathbf{T}_D^{BC}\mathbf{R}_U^{AB}[\mathbf{I} - \mathbf{R}_D^{BC}\mathbf{R}_U^{AB}]^{-1}\mathbf{T}_U^{BC}$$

$$\mathbf{T}_U^{AC} = \mathbf{T}_U^{AB}[\mathbf{I} - \mathbf{R}_D^{BC}\mathbf{R}_U^{AB}]^{-1}\mathbf{T}_U^{BC}. \quad\ldots\ldots\ldots\ldots\ldots\ldots\ldots (2.39)$$

Note that we obtain our solution in the frequency-ray parameter domain. An inverse Fourier transform derives the $(\tau\text{-}p)$ seismograms. Synthetics in the offset-time domain can be obtained by transformation of the (ω, p) or $(\tau\text{-}p)$ seismograms.

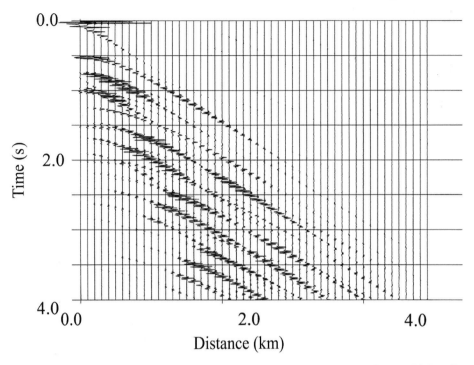

P- and S- wave Primaries and all surface multiples

Time (s)

Distance (km)

Fig. 2.10d—Synthetic seismograms including P- and S-waves and all surface multiples. Note that even for a simple layer over a half-space model, the seismograms are very complicated.

A set of synthetic seismograms that includes all the surface multiples and converted waves for a model shown in Fig. 2.10a is displayed in Fig. 2.10d. We note that even for a simple model such as a layer over a half space, the response is fairly complicated.

2.9 Seismogram Synthesis in Laterally Heterogeneous Media

The basic assumption employed in reflectivity modeling, described in the previous section, is that elastic properties vary along only one coordinate axis (i.e., depth). Although some techniques have recently been proposed to modify the reflectivity approach to include lateral heterogeneity by allowing for the coupling of plane waves with the spatial spectrum of the medium, their application is far from practical. Several methods have been proposed to model seismic wave propagation in heterogeneous media, which can be grouped into two principal categories: analytic solutions that are generally asymptotic and numerical solutions. It is not possible to derive an analytic solution of the seismic wave equation for a general heterogeneous medium, but an analytic solution can be derived when the model is simplified. Asymptotic ray theory (ART) assumes that the medium is smoothly varying and derives solutions for a ray path that is valid at high frequencies. ART is the most commonly employed seismic modeling tool in heterogeneous media; most of what we know today about the Earth's deep interior has been derived by ray theoretical modeling. The algorithm is fast and is therefore very useful for interpreting seismograms. Ray solutions are, how-

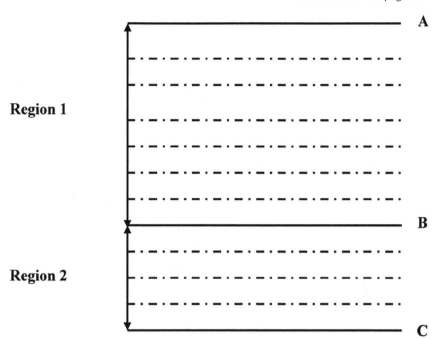

Fig. 2.11—A recursive scheme is used to compute the response of the zone AC, when the responses of zones AB and BC are known.

ever, far from complete, because there are always too many rays to count, and the solutions break down for complex geologic models in which the waves generate shadow zones and caustics.

Numerical methods for solving the wave equation include the methods of finite differences (FD) and finite elements (FE), of which the FD has been the most popular in the seismology community. It is a direct and straightforward approach to solving the wave equation. Its salient features include the use of basic equations, initial conditions, and some boundary conditions, and it requires minimal analytic effort. It is straightforward to write an FD computer code. However, the method can make intensive demands on computer memory and CPU. Therefore, the size and complexity of the problem that can be solved by FD is limited by the capabilities of available computers.

Finite-differencing is a method of approximating (numerical evaluation) derivatives based on a Taylor series expansion of a function. For example, a function $\phi(x)$ can be expanded in a Taylor series as

$$\phi(x \pm \Delta x) = \phi(x) \pm \frac{\partial \phi}{\partial x} \Delta x + \frac{1}{2} \partial_x^2 \phi (\Delta x)^2 \pm \frac{1}{6} \partial_x^3 \phi (\Delta x)^3 + - -. \qquad \ldots \ldots (2.40)$$

We define the backward-differencing formula by

$$\phi'(x) = \frac{\partial \phi}{\partial x} = \frac{1}{\Delta x} [\phi(x - \Delta x) - \phi(x)]. \qquad \ldots \ldots \ldots \ldots \ldots \ldots (2.41)$$

and the forward-differencing formula by

$$\phi'(x) = \frac{\partial \phi}{\partial x} = \frac{1}{\Delta x} [\phi(x) - \phi(x + \Delta x)]. \qquad \dots \dots \dots \dots \dots \dots (2.42)$$

The truncation error for both formulae has a leading term proportional to Δx. On the other hand, the central difference formula is defined by

$$\phi'(x) = \frac{\partial \phi}{\partial x} = \frac{1}{2\Delta x} [\phi(x + \Delta x) - \phi(x - \Delta x)], \qquad \dots \dots \dots \dots \dots (2.43)$$

which has a leading error of the order of $(\Delta x)^2$.

Similar expressions can be derived for the second derivatives. By adding a Taylor series for $\phi(x + \Delta x)$ and $\phi(x - \Delta x)$, we find that

$$\frac{\partial^2 \phi}{\partial x^2} = \frac{\phi(x + \Delta x) - 2\phi(x) + \phi(x - \Delta x)}{\Delta x^2}. \qquad \dots \dots \dots \dots \dots (2.44)$$

A straightforward finite differencing approach involves expressing all the spatial and temporal derivatives appearing in the second order elastic wave equation given by Eq. 2.6 in terms of the finite difference equations given above. By rearranging terms in the difference equation, we can solve for displacement at a future time at all locations in the subsurface using the displacements known at all grid points at two past times (Kelly *et al.* 1976). Velocity and density models are discretized on rectangular grids, and a source is injected at a point (or a box around it); the displacement at all future times and at all grid points can now be computed using the finite difference approximations of the wave equation. It is fairly easy to write a computer code to implement the finite difference algorithm. However, for practical applications, we need to choose the spatial grid-spacing and time-step size carefully to avoid numerical artifacts. This is because the choices of the spatial grids and time-step size are dictated by the *grid dispersion criterion* and the *stability criterion*. The grid dispersion criterion requires the spatial grid spacing

$$\Delta x < n \frac{\alpha_{min}}{\omega_{max}},$$

where Δx is the spatial grid interval, α_{min} is the minimum velocity, ω_{max} is the highest frequency to be modeled, and the value of n depends on the differencing scheme. Similarly, the stability criterion is given by

$$\Delta t < m \frac{\Delta x}{\alpha_{max}}.$$

Thus we notice that the grid spacing Δx and the time-step increment Δt are determined by the limiting values of the velocities and the highest frequency; m depends on the differencing scheme. These conditions are generally satisfied globally.

One additional difficulty with FD approaches is that the edges of the model act as perfect reflectors, and desired reflections are often contaminated with the boundary reflection effects. To avoid this, we need to either expand the model or implement an *absorbing boundary condition* at the edges that attempts to minimize edge reflections. The FD formulation of the second-order wave equation as discussed here also requires that the boundary conditions at the fluid-solid boundary be coded explicitly.

Most FD codes commonly used in the seismic community make use of a first-order system of the equation of motion for particle velocity and stress (Eq. 2.4 and 2.5) with a staggered grid approach (Levander 1988). This approach requires fewer grid points per wavelength to avoid numerical dispersion and requires no special treatment for fluid-solid boundaries. A pseudo-FD code is shown in **Fig. 2.12**. An example of FD synthetic seismograms for a 2D laterally-varying model (SEG-EAGE salt model) is shown in **Fig. 2.13**. In a laterally heterogeneous medium, scattering, focusing, and defocusing effects may also cause amplitude variations.

2.10 Viscoelasticity
In all the preceding discussions in this chapter, we assumed the rock layers to be perfectly elastic and to obey Hooke's law. In reality, however, all Earth materials deviate from the standard Hooke's law in many ways, and studying the effects of anelasticity becomes very

---

- Read input parameters—number of grid points in three dimensions, source location, peak frequency, receiver geometry, etc.

- Read velocity and density files representing these parameters in a 2D/3D grid.

- Read input parameters—source location, peak frequency, receiver geometry, etc.

- Generate wavelet.

Loop over timestep

- Inject source.

- Update particle velocities using FD scheme.

- Update stresses using FD scheme.

- Apply absorbing boundary condition to attenuate edge reflections.

- Output seismograms.

End loop

Fig. 2.12—A pseudocode for seismic wave propagation simulation using finite differences.

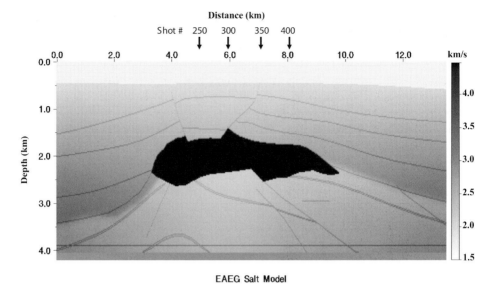

Fig. 2.13a—SEG-EAGE salt model.

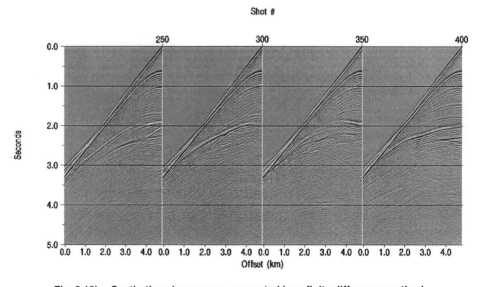

Fig. 2.13b—Synthetic seismograms computed by a finite difference method.

important in many situations. Generally, Earth materials deviate from elastic behavior by exhibiting viscous-like as well as elastic characteristics. Viscoelastic materials are those for which the relationship between stress and strain is time-dependent and possesses the following characteristics:

- Creeping—if the stress is held constant, the strain increases with time.
- Relaxation—if the strain is held constant, the stress decreases with time.

- The effective stiffness (reciprocal of elastic coefficient) depends on the rate of application of the external force.
- The seismic waves experience attenuation.

When the material is not purely elastic, the seismic waves experience a loss of energy during propagation in addition to the losses caused by spherical divergence, scattering, diffraction, reflection, and transmission. The behavior is explained using a stress-strain relation via a complex modulus (e.g., Aki and Richards 2002). To understand this behavior, let us consider a plane wave propagating in an elastic homogeneous medium, the amplitude of which is given by

$$A(x,t) = A_0 \exp[i(kx - \omega t)], \qquad \dots\dots\dots\dots\dots\dots\dots (2.45)$$

where A_0 is the maximum amplitude, k is the wave number, and ω is the frequency. All these quantities are real. In an attenuating medium, we assume the elastic modulus or the velocity or the wave number to be complex. In the case of a complex wave number, we have

$$k = k_r + ik_i, \qquad \dots\dots\dots\dots\dots\dots\dots\dots\dots (2.46)$$

where k_r is the real part of the wave number given by

$$k_r = \frac{\omega}{\alpha(\omega)}, \qquad \dots\dots\dots\dots\dots\dots\dots\dots\dots (2.47)$$

and k_i is the plane-wave attenuation coefficient, which has the units of reciprocal of length. Thus the plane-wave equation becomes

$$A(x,t) = A_0 \exp(-k_i x)\exp i(k_r x - \omega t), \qquad \dots\dots\dots\dots\dots (2.48)$$

clearly demonstrating that the plane-wave amplitude decays exponentially away from the source determined by the coefficient k_i.

Using this definition, the attenuation coefficient k_i can be expressed in terms of the amplitude of the wave at two different positions x_1 and x_2 as

$$k_i = \frac{1}{(x_2 - x_1)} \ln\left(\frac{A(x_1)}{A(x_2)}\right). \qquad \dots\dots\dots\dots\dots\dots\dots (2.49)$$

It is also defined in units of dB/unit length as follows:

$$k_i = \frac{1}{x_2 - x_1} 20\ln\left(\frac{A(x_1)}{A(x_2)}\right). \qquad \dots\dots\dots\dots\dots\dots (2.50)$$

Another measure of attenuation is called Q or the *quality factor* and is used most commonly. O'Connell and Budiansky (1978) discussed in detail the various definitions of the quality factor. The internal loss of energy during wave propagation gives the intrinsic Q value and is in excess of all other losses encountered during wave propagation, as discussed earlier in this chapter.

2.11 Factors Affecting Seismic Amplitudes

In this chapter we have discussed the fundamental principles behind seismic wave propagation, the assumptions made in the theoretical developments, and algorithms for generating the seismic response of the Earth. It is obvious that the travel times are affected by velocity alone and Fermat's principle determines ray paths in complex geologic models. Travel time is the most important seismic attribute and the velocities determined from modeling travel times are used to obtain meaningful images and subsurface configurations of reservoirs. Seismic amplitudes, on the other hand, depend on a variety of factors and are useful for obtaining detailed maps of rock properties. Based on our discussions, we can identify the following factors affecting seismic amplitudes:

- Source directivity
- Spreading
- Reflection/transmission at and across material discontinuities
- Attenuation
- Focusing/defocusing effects
- Scattering
- Receiver directivity.

All these factors, except the source and receiver directivities, are caused by propagation effects, and need to be carefully examined prior to making geologic interpretations based on seismic wave amplitudes.

Chapter 3

Inverse Theory

The word *inversion* has been used to mean a wide variety of things in different applications in science and engineering. The practitioners are often confused about the meaning of inversion, which has resulted in a large degree of skepticism about its usefulness. In general, what one considers inversion depends on one's background, such as engineering, mathematics, or geophysics. Stolt (1989) describes the state of affairs as follows:

"What I am doing is inversion—I don't know what you are doing."
"What they're doing is inversion, so it can't be of practical use."

Occasionally, segments of the inversion community have promoted an elitist image, which has encouraged this response by those outside that community. This is also the reason for many young graduate students to shy away from learning inverse theory. Much of the theoretical developments are based on vector spaces, linear algebra, and statistics, and the details of implementation are generally masked by mathematical jargon. Several textbooks on geophysical inversion have recently been published (e.g., Menke 1984; Tarantola 1987; Sen and Stoffa 1995), each focused on different aspects of inversion. In this chapter, we attempt to describe the fundamental principles in a simple manner.

An important task of the physical sciences is to make inferences about physical parameters from data. We go through the process of guessing, computing, and comparing to draw inferences. Inversion involves very similar (if not identical) steps. Formally, we can define inverse theory as a set of mathematical techniques for reducing data to obtain useful information about the physical world on the basis of inferences drawn from observations (Menke 1984).

In general, laws of physics provide the means for computing the data values, given a model. This is called the *forward model*. The problem of generating theoretical data given a model is termed a *forward problem*. For example, elastic wave theory formulation for computing synthetic seismograms, as described in Chapter 2, constitutes the forward problem. Given a set of parameters describing a model (also called *model parameters*), forward modeling is employed to compute theoretical data (**Fig. 3.1**). In the case of seismic wave propagation, the Earth model parameters consist of elastic properties such as the compressional wave velocity, shear-wave velocity, and density at different points in the subsurface. Synthetic seismograms can be generated using a reflectivity or a finite-difference algorithm, for example.

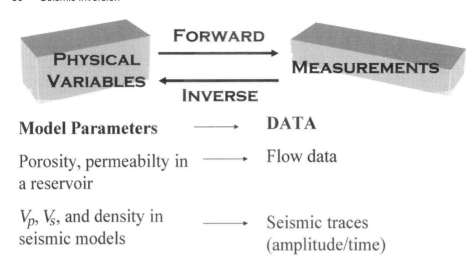

Fig. 3.1—A schematic diagram describing forward and inverse problems: forward modeling involves mapping models to data, while inversion involves estimation of model parameters from observations.

3.1 Inversion

Just as the forward problem evaluates theoretical data, the goal of inversion is to derive a description of Earth model parameters from recorded observations. In principle, it appears straightforward in that one needs to employ "reverse physics" to derive Earth model parameters. In practice, such *direct inversion* algorithms are highly unstable, and model parameters are generally estimated by "trial and error" matching of observed data with synthetic data by solving the forward problem for a large number of trial Earth models. Formally, this is done in an automated manner by using the techniques of *optimization,* in which we define a mathematical function (called *cost function, misfit function,* or *objective function*) that measures the misfit between the observed and synthetic data and seek optimal model parameters for which the data fit is optimal measured by a minimum of the objective function **(Fig. 3.2)**. Such an approach is often called a *model-based inversion,* and optimization plays an important role in it. The goal of inversion, however, is far beyond just finding a "best-fit" model. The data are often noisy, the forward modeling is inexact, and there may not be sufficient data. For these reasons, most inverse problems have nonunique solutions. In other words, multiple models often can fit the measured data almost equally well. Therefore, it becomes important not only to find a model but also to estimate the uncertainty **(Fig. 3.3)**. We will focus primarily on the model-based inversion algorithms and describe some of these aspects in details in the following sections.

Examples of some inverse problems are listed below.

Geophysics
- Gravity and magnetic anomalies.
- Seismic tomography and earthquake location.
- Deconvolution.
- Amplitude variation with offset.

ITERATIVE FORWARD MODELING

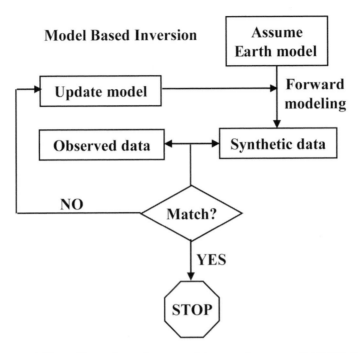

Fig. 3.2—In a model-based inversion scheme, model parameters are estimated by iterative forward modeling.

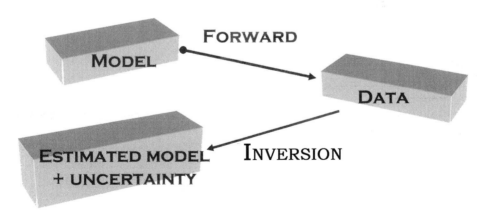

Fig. 3.3—Because of inaccuracies in the data and inadequate forward modeling, an inversion requires estimation of model parameters as well as uncertainties.

Reservoir Engineering
- History matching.
- Pressure transient analysis.

Medical Imaging
- Transmission computed tomography (CT).
- Emission computer tomography (SPECT—single photon emission computerized tomography; PET—positron emission tomography).
- Ultrasound computed tomography.
- Electric source imaging (ESI).
- Magnetic resonance imaging (MRI).

3.2 Model-Based Inversion Methods

In this approach, synthetic data are generated for an assumed model and compared with observed data. If the match between the observed and the synthetic data is acceptable, the model is accepted as the solution. Otherwise, the model is changed, and the synthetics are recomputed and recompared with the observations. This iterative forward modeling procedure is repeated until an acceptable match is obtained between the data and the synthetics. Thus, in this approach, inversion is viewed as an *optimization* process in which a model is sought that best explains the observations. The fit (measure of agreement) or misfit (measure of disagreement) between the observed and synthetic data is used as a measure of acceptability of an Earth model. Optimization methods vary based on their method of search for the optimum model. The simplest of the optimization methods is one that assumes a linear relationship between data and model, in which case an answer can be obtained in one step.

In geophysics, we always deal with discrete data; thus, it is convenient to represent the data as a column vector of the type

$$\mathbf{d} = [d_1, d_2, d_3, \ldots, d_{ND}]^T, \quad\quad\quad\quad\quad\quad (3.1)$$

where ND is the number of data points and T is a matrix transpose. For example, in seismology, each data value is a time sample in a seismogram and $ND = NT \cdot NTR$, where NT is the number of samples in a seismogram and NTR is the number of seismograms.

Similarly, a column vector can also represent an Earth model

$$\mathbf{m} = [m_1, m_2, m_3, \ldots, m_{NM}]^T, \quad\quad\quad\quad\quad\quad (3.2)$$

where NM is the total number of model parameters. For example, for a 1D acoustic problem in seismology as described in Chapter 2, the model parameters can be used to represent compressional wave velocity (α_i), density (ρ_i), and thickness (h_i) of all the layers.

Synthetic data are generated by forward calculation using a model vector $\mathbf{m}$. Thus the synthetic data vector $\mathbf{d}_{syn}$ can be computed by application of a forward modeling operator, g, to the model vector $\mathbf{m}$; i.e.,

$$\mathbf{d}_{syn} = g(\mathbf{m}). \quad\quad\quad\quad\quad\quad (3.3)$$

In most geophysical problems, the forward modeling operator g is a nonlinear operator, as described in Chapter 2.

The inverse problem now reduces to determining the model(s) that minimize the misfit between the observed and the synthetic data. The misfit function is also called the *objective function, cost function, error function, energy function*, etc., and is usually given by a suitably-defined *norm*. The error vector **e** is given by

$$\mathbf{e} = \mathbf{d}_{obs} - \mathbf{d}_{syn} = \mathbf{d}_{obs} - g(\mathbf{m}). \qquad \dots\dots\dots\dots\dots\dots\dots\dots (3.4)$$

A general norm, L_p (e.g., Menke 1984), is defined as

$$L_p \text{ Norm} \quad : \quad \|\mathbf{e}\|_p = \left[\sum_{i=1}^{ND} |e_i|^p \right]^{1/p}, \qquad \dots\dots\dots\dots\dots\dots\dots (3.5)$$

where ND is the number of data points. Thus, the commonly-used L_2 norm is given by

$$L_2 \text{ Norm} \quad : \quad \|\mathbf{e}\|_2 = \left[\sum_{i=1}^{ND} |e_i|^2 \right]^{1/2}, \qquad \dots\dots\dots\dots\dots\dots\dots (3.6)$$

or, in vector notation, we have

$$L_2 \text{ Norm} \quad : \quad \|\mathbf{e}\|_2 = [(\mathbf{d}_{obs} - g(\mathbf{m})^T)(\mathbf{d}_{obs} - g(\mathbf{m}))]^{1/2}. \qquad \dots\dots\dots\dots\dots (3.7)$$

Usually the above equation is divided by the number of observation points ND, in which case it reduces to the well-known root mean square (RMS) error. Higher norms give larger weights to large elements of **e**. In many geophysical applications the L_2 norm has been used; use of other types of norms (e.g., L_1) can also be found in the geophysical literature.

At this point, we note the following features of model-based geophysical inverse problems:

- Because *g* in Eq. 3.3 is in general a nonlinear operator, the error function (as a function of the model) is complicated in shape. Thus, it can be expected to have multiple minima of varying heights. Only if *g* actually is or can be approximated by a linear operator will the error function become quadratic with respect to perturbations in the model parameters, and only then will it have only one well-defined minimum.
- For many geophysical applications, generation of synthetic data is a highly time-consuming task, even on fast computers.
- The search space of model parameters is generally very large.
- Definition of an objective function simply as a difference between observed and synthetic data is often not adequate, because the data and model have varying magnitudes and dimensions, and multiple solutions are possible.

3.2.1 Classification of Inverse Problems. Inverse problems are often classified based on the nature of the relationship between the data and the model and the behavior of the objective function, into the following categories: linear, weakly nonlinear, quasilinear, and nonlinear.

Linear Problems. The relationship between data and model is described generally by a nonlinear operation described in Eq. 3.3, in which the forward modeling operator *g* is a nonlinear operator. In many applications, the forward modeling operator can be approximated by a linear operator or a matrix $\mathbf{G}$ such that

$$\mathbf{d}_{syn} = \mathbf{Gm}. \quad \dots\dots\dots\dots\dots\dots\dots\dots\dots\dots\dots\dots\dots\dots (3.8)$$

Such problems are referred to as *linear inverse* problems. Note that approximating a non-linear forward modeling operator with a nonlinear operator results in several consequences. Nonetheless, for many applications, a linear inversion has been found to be adequate to learn the nature of the system. Much of classical inverse theory is based on linear inverse problems, and many concepts and techniques are easily applicable to more general problems.

The solution of Eq. 3.8 can be written as

$$\mathbf{m} = \mathbf{G}^{-1}\mathbf{d}_{obs}, \quad \dots\dots\dots\dots\dots\dots\dots\dots\dots\dots\dots\dots\dots\dots (3.9)$$

if $\mathbf{G}^{-1}$ exists. Generally the solution is approximated by the following equation:

$$\mathbf{m}_{est} \approx \mathbf{G}^g\mathbf{d}_{obs}, \quad \dots\dots\dots\dots\dots\dots\dots\dots\dots\dots\dots\dots\dots (3.10)$$

where $\mathbf{G}^g$ is called a *generalized inverse* and $\mathbf{m}_{est}$ is the estimated model vector. The solution can be obtained in one step if $\mathbf{G}^g$ can be constructed, which is not a trivial task.

Weakly Nonlinear Problems. In a weakly nonlinear formulation, the computation of data for a given model $\mathbf{m}$ is done simply by linear perturbation to data for a reference model $\mathbf{m}_0$, which is assumed to be very close to $\mathbf{m}$; i.e.,

$$\mathbf{d} = g(\mathbf{m}) \approx g(\mathbf{m}_0) + \frac{\partial g}{\partial \mathbf{m}}(\mathbf{m} - \mathbf{m}_0) + \dots = \mathbf{d}_0 + \mathbf{G}_0\Delta\mathbf{m}, \quad \dots\dots\dots\dots (3.11)$$

where $\mathbf{G}_0$ is a matrix containing the partial derivatives of the data with respect to model parameters. Equation 3.11 can be rewritten as

$$\Delta\mathbf{d} \approx \mathbf{G}_0\Delta\mathbf{m}, \quad \dots\dots\dots\dots\dots\dots\dots\dots\dots\dots\dots\dots\dots (3.12)$$

where $\Delta\mathbf{d}$ is the data residual and $\Delta\mathbf{m}$ is the model perturbation. Thus, in Eq. 3.11, we have a linear system very similar to Eq. 3.8, and the solution (model perturbation) can be obtained using techniques for solving linear inverse problems.

Quasilinear Problems. Unlike the weakly nonlinear case, in a quasilinear problem (they are indeed commonly referred to as nonlinear problems), rather than linearizing the forward model, the objective function is linearized. In other words, the objective function is assumed to be linear in the close vicinity of a model in consideration. That is, we write

$$e(\mathbf{m} + \Delta\mathbf{m}) \approx e(\mathbf{m}) + \Delta\mathbf{m}^T\nabla e(\mathbf{m}), \quad \dots\dots\dots\dots\dots\dots (3.13)$$

where $\nabla e(\mathbf{m})$ contains the partial derivative of error with respect to model parameter. With

$$\mathbf{F} = \nabla e(\mathbf{m}), \quad \dots\dots\dots\dots\dots\dots\dots\dots\dots\dots\dots\dots\dots (3.14)$$

we have the same form as the weakly nonlinear case. Note that often the objective function is approximated as piecewise linear, and the solution is obtained in an iterative fashion. Local optimization methods (discussed later in the chapter) are generally used for this purpose.

Nonlinear Problem. In a nonlinear inverse problem, the forward modeling is treated as a nonlinear functional, and the objective function is also retained in its true form. No assumption is made on the behavior of the objective functions.

3.3 Examples of Some Geophysical Inverse Problems

3.3.1 Inversion of Gravity Anomalies.

Gravity anomalies are caused by density contrasts of the subsurface rocks. Unlike a seismic exploration experiment, gravity is a natural experiment where the observations are the measurements of gravity fields at several points on the surface (occasionally measurements are made in boreholes), and the model parameters are the shape of the causative body and the density distributions. Assuming that the function $d(x, y, z)$ represents a gravity anomaly measured at a point (x, y, z) and that $\rho(x, y, z)$ represents density distribution, we can write the following equation for computation of gravity anomalies:

$$d(x,y,z) = \int_{-\infty}^{\infty} \int_{-\infty}^{\infty} \int_{z_1}^{z_2} \rho \, \frac{G}{r^3} \, dx' dy' dz', \quad \dots\dots\dots\dots\dots\dots\dots (3.15)$$

where G is the gravitational constant and the integral is carried over the entire causative body. We assume here that the body is extended to infinity in x and y directions while in depth it is contained within levels z_1 and z_2, and r is the distance between a point in the body and the observation point.

There are two approaches to formulating the inverse problem. In the first approach, we assume that the shape of the body is known and that inversion attempts to estimate density distribution. In such a situation, the Kernel function in Eq. 3.15 is independent of the coordinates of the body, which cause nonlinearity in Eq. 3.15. For a discrete problem, Eq. 3.15 reduces to the following simple linear form:

$$\mathbf{d} = \mathbf{H}\rho, \quad \dots\dots\dots\dots\dots\dots\dots\dots\dots\dots\dots\dots\dots\dots\dots (3.16)$$

where $\mathbf{d}$ is a vector of observed gravity, ρ is a model vector containing the unknown density distribution, and $\mathbf{H}$ is the Kernel matrix (a forward modeling operator).

In the second approach, we assume that the density is known and attempt to estimate the shape of the body. The computation involves solution of a nonlinear integral equation, because the unknown quantities occur within the Kernel function. This results in a nonlinear inverse problem.

The gravity inverse problem is an interesting practical example from geophysics in which the problem can be mapped into a linear or a nonlinear inverse problem, depending on the objectives.

3.3.2 Inversion of Seismic Amplitude.

In Chapter 2, we discussed the contribution of several factors to variations in seismic wave amplitude with angle or offset. One of the most important factors affecting the seismic amplitude is the plane-wave reflection coefficient that determines the partitioning of seismic energy separating two media in welded contact. The equation for modeling reflection coefficient is a nonlinear function of model parameters such as the P-wave and S-wave velocities and the densities of the two media. We also discussed a simpler linear form (Eq. 2.27) given by the following equation:

$$Rpp(p) \cong \frac{1}{2}(1 - 4\beta^2 p^2)\frac{\Delta\rho}{\rho} + \frac{1}{2\cos^2 i}\frac{\Delta\alpha}{\alpha} - 4\beta^2 p^2\frac{\Delta\beta}{\beta}.$$

For discrete data (N rays and one reflector), the above equation can be written in the following form:

$$\begin{bmatrix} Rpp(p_1) \\ Rpp(p_2) \\ \vdots \\ Rpp(p_N) \end{bmatrix} = \begin{bmatrix} \frac{1}{2}(1 - 4\beta^2 p_1^2) & \frac{1}{2\cos^2 i_1} & -4\beta^2 p_1^2 \\ \frac{1}{2}(1 - 4\beta^2 p_2^2) & \frac{1}{2\cos^2 i_2} & -4\beta^2 p_2^2 \\ \vdots & \vdots & \\ \frac{1}{2}(1 - 4\beta^2 p_N^2) & \frac{1}{2\cos^2 i_N} & -4\beta^2 p_N^2 \end{bmatrix} \begin{bmatrix} \frac{\Delta\rho}{\rho} \\ \frac{\Delta\beta}{\beta} \\ \frac{\Delta\alpha}{\alpha} \end{bmatrix}, \quad \dots\dots\dots (3.17)$$

where the model parameters are the contrasts in material properties and p is the ray-parameter (or angle).

3.4 Solution of Linear Inverse Problems

Once we recognize that a problem is linear or linearizable with respect to a reference model, equations of the form (3.8) and (3.12) can be written down. In Eq. 3.8, the matrix **G** represents a forward modeling operator and is known from the forward relation. In Eq. 3.12, the elements of the $\mathbf{G}_0$ matrix are the partial derivatives of data with respect to model parameters (computed for the reference model), which need to be calculated. Once **G** is known, we can solve for the model vector by applying the inverse operator for **G**, $\mathbf{G}^{-1}$ (if it exists) to the data vector. Similarly, applying the inverse of $\mathbf{G}_0$ to the data residual vector, we can obtain an update to the model perturbation vector. This, however, may not be a trivial task; much of the subject of linear inverse theory deals with issues related to obtaining estimates of the inverse operators for the solution of linear inverse problems. The solution methods, the nature of the solution obtained, and the conditions under which solutions may be obtained are but a few of the topics for which geophysical inverse theorists were able to introduce many new ideas and concepts.

Formally, the inverse solution of Eq. 3.8 can be written as

$$\mathbf{m} = \mathbf{G}^{-1}\mathbf{d}_{syn},$$

where $\mathbf{G}^{-1}$ is the inverse operator of **G**. This leads us to the following issues related to the problem of finding solutions.

3.4.1 Existence. The existence implies that of a solution **m** or of the inverse operator $\mathbf{G}^{-1}$.

3.4.2 Uniqueness. A solution is said to be unique if changing a model from $\mathbf{m}_1$ to $\mathbf{m}_2$, the data also change from $\mathbf{d}_1$ to $\mathbf{d}_2$ such that $\mathbf{d}_1 \neq \mathbf{d}_2$ (i.e., the operator $\mathbf{G}$ is *injective*). Otherwise, several models will explain the data equally well, and we obtain nonunique solutions. Some of the reasons for the nonuniqueness of the results for geophysical inverse problems are:

- For problems dealing with Earth parameters, the infinite dimensionality of the model space is usually formulated into a discrete finite problem resulting in inherent nonuniqueness. Material properties in the real Earth are continuous functions of the spatial coordinates. In inversion, attempts are made to derive these functions from measurements of a finite number of data points, causing nonuniqueness in results.
- Uniqueness is also related to the problem of *identifiability* of the model by the data. For example, in seismic problems, the information on seismic wave velocity is contained in the moveout (travel time as a function of offset) in the data. Therefore, the use of only near-offset traces does not reveal much information on velocity. This means that a suite of velocities will explain the data equally well, resulting in nonunique estimates of velocities. In seismic tomography problems, the regions of the Earth with no ray coverage cannot be resolved from the data, and slowness estimates for these regions will result in nonunique values. Thus, resolution and uniqueness are closely related. Also in many situations, we may have a clear tradeoff between model parameters.

3.4.3 Stability. Stability indicates how small errors in the data propagate into the model. A stable solution is insensitive to small errors in the data values. Instability may induce nonuniqueness, which enhances the inherent nonuniqueness in geophysical applications.

3.4.4 Robustness. Robustness indicates the level of insensitivity with respect to a small number of large errors (outliers) in the data.

Inverse problems that do not possess uniqueness and stability are called *ill-posed* inverse problems. Otherwise, the inverse problem is called *well-posed*. Techniques known as *regularization* can be applied to ill-posed problems to restore well-posedness.

Because many geophysical inversions result in nonunique solutions, the objective of inversion is to first find a solution (or solutions) and to represent the degree of nonuniqueness of the solution in a quantitative manner. During the process, attempts should be made to reduce nonuniqueness, if possible, and/or explain it in terms of data errors and the physics of the forward problem.

3.4.5 Method of Least Squares. Recall that for a linear inverse problem the L_2-norm error function is given by

$$E(\mathbf{m}) = \mathbf{e}^T \mathbf{e} = (\mathbf{d}_{\text{obs}} - \mathbf{Gm})^T (\mathbf{d}_{\text{obs}} - \mathbf{Gm}). \qquad \ldots\ldots\ldots\ldots\ldots\ldots\ldots (3.18)$$

This can be solved by locating the minimum of $E(\mathbf{m})$ where the derivative of $E(\mathbf{m})$ (with respect to $\mathbf{m}$) vanishes. That is,

$$\frac{\partial E(\mathbf{m})}{\partial \mathbf{m}} = 0, \qquad \ldots\ldots\ldots\ldots\ldots\ldots\ldots\ldots\ldots\ldots\ldots\ldots\ldots (3.19)$$

which gives

$$\mathbf{G}^T\mathbf{Gm} - \mathbf{G}^T\mathbf{d} = 0, \quad \dots\dots\dots\dots\dots\dots\dots\dots\dots\dots\dots\dots\dots (3.20)$$

where $\mathbf{0}$ is a null vector. This yields

$$\mathbf{m}_{est} = [\mathbf{G}^T\mathbf{G}]^{-1}\mathbf{G}^T\mathbf{d}, \quad \dots\dots\dots\dots\dots\dots\dots\dots\dots\dots\dots\dots (3.21)$$

assuming that $[\mathbf{G}^T\mathbf{G}]^{-1}$ exists, where $\mathbf{m}_{est}$ is the least squares estimate of the model. Thus, from knowledge of the $\mathbf{G}$ matrix, one can derive estimates of the model parameters in one step. G is an (*NDXNM*) matrix, and $\mathbf{G}^T\mathbf{G}$ is a symmetric square matrix with (*NMXNM*) elements. Whether the least squares solution exists or not depends on $[\mathbf{G}^T\mathbf{G}]^{-1}$, which, in turn, depends on how much information the data vector $\mathbf{d}$ possesses on the model parameters. The matrix $[\mathbf{G}^T\mathbf{G}]^{-1}\mathbf{G}^T$ operates on the data to derive model parameters [i.e., it inverts the system of linear equations (Eq. 3.8)]. We will represent this matrix with the symbol $\mathbf{G}^g$, such that

$$\mathbf{G}^g = [\mathbf{G}^T\mathbf{G}]^{-1}\mathbf{G}^T. \quad \dots\dots\dots\dots\dots\dots\dots\dots\dots\dots\dots\dots (3.22)$$

One of the necessary conditions for the system of linear equations $\mathbf{Gm} = \mathbf{d}$ to have one unique solution is that there must be as many equations as the number of unknown model parameters. This is the situation when we have as many data as the number of model parameters (*ND* = *NM*) such that these data contain information on all the model parameters. Such a problem is called an *even-determined* problem. In an *underdetermined* problem *NM* > *ND* (i.e., there are fewer data than the number of model parameters). This would usually be the case with geophysical inversion if we attempt to estimate Earth model parameters that are continuous (or infinite) from a finite set of measurements. The problem can be reduced to an *even-determined* or even an *overdetermined* case by discretization, which reduces the number of model parameters. For example, instead of solving for the Earth parameters as a continuous function of spatial coordinates, we may divide the Earth model into a set of discrete layers (based on independent information). A problem may be *mixed-determined* as well, in which data may contain complete information on some model parameters and none on the others. Such a situation may arise in seismic tomography problems when there is complete ray coverage in some of the blocks, while other blocks may be completely devoid of any ray coverage. Thus, a problem can be *underdetermined* even in cases with *ND* > *NM*. In the under-determined case, several solutions exist, and the model parameter estimates will be nonunique. In situations in which we have more data than the number of model parameters and the data contain information on all model parameters, the problem is said to be *overdetermined*. Clearly, in such a situation, the linear system of equations cannot find an answer that fits all data points unless they all lie on a straight line. In this case, the best estimate can be obtained in the least squares sense, meaning that the error is a nonzero smallest value.

3.4.6 Maximum Likelihood Methods. Geophysical data are often contaminated with noise, and therefore, every data point may be uncertain, the degree of uncertainty being different for different data points. It is also possible that the data points may influence each other (i.e., they may be correlated). Thus, each data point can be considered a random variable such that the vector $\mathbf{d}$ now represents a vector of random variables. If each of the data variables is assumed to be Gaussian distributed, their joint distribution is given by

$$p(\mathbf{d}) \propto \exp\left[-\frac{1}{2}(\mathbf{d} - \langle\mathbf{d}\rangle)^T \mathbf{C}_d^{-1}(\mathbf{d} - \langle\mathbf{d}\rangle) \right], \quad \dots\dots\dots\dots\dots\dots\dots (3.23)$$

where $\langle\mathbf{d}\rangle$ and $\mathbf{C}_d$ are the mean data and data covariance matrix, respectively. $P(\mathbf{d})$ gives the probability of observation of the data values. For a linear inverse problem, we can write the above probability as

$$p(\mathbf{d}) \propto \exp\left[-\frac{1}{2}(\mathbf{d} - \mathbf{Gm})^T \mathbf{C}_d^{-1}(\mathbf{d} - \mathbf{Gm}) \right]. \quad \dots\dots\dots\dots\dots\dots (3.24)$$

This means that the application of $\mathbf{G}$ on the estimated model results in the mean values of data and $\mathbf{d}$ is a single realization of a random data vector. Thus, the optimum values of the model parameters are the ones that maximize the probability of measured data, given the uncertainties in the data. This procedure is an example of the *maximum likelihood method* (MLM).

For numerical implementation, this method is not very different from the least squares method (for the case of Gaussian distributed data) described in the previous section. The maximum probability occurs when the argument of the exponential is minimum. Thus, instead of minimizing the error function given by Eq. 3.18, we minimize the following error function:

$$E_1(\mathbf{m}) = (\mathbf{d}_{obs} - \mathbf{Gm})^T \mathbf{C}_d^{-1}(\mathbf{d}_{obs} - \mathbf{Gm}), \quad \dots\dots\dots\dots\dots\dots (3.25)$$

which can again be solved by the least squares method. By finding the minimum where the derivative of the error with respect to the model parameters is zero, we can derive an equation analogous to Eq. 3.21. This form reduces to the case of simple least squares when $\mathbf{C}_d$ is an identity matrix. In the general case, however, each data residual is weighted inversely with the corresponding element of the data covariance matrix. This means that for a diagonal $\mathbf{C}_d$, the data with large uncertainty will have a relatively smaller contribution to the solution, while those with small error will have a relatively larger contribution. The process attempts to fit the reliable data values better than the unreliable ones.

Even without knowledge of $\mathbf{C}_d$ as required in the maximum likelihood method, it may be desirable to use a weighting matrix, $\mathbf{W}_d$, as in the following definition of the error function:

$$E_2(\mathbf{m}) = (\mathbf{d}_{obs} - \mathbf{Gm})^T \mathbf{W}_d^{-1}(\mathbf{d}_{obs} - \mathbf{Gm}). \quad \dots\dots\dots\dots\dots\dots (3.26)$$

The need for a weighting matrix can occur because of assumed data uncertainties or because the elements of the data vector may have different dimensions or ranges of values. In the conventional definition of the error function, all the data values have equal weight, and therefore, the data with large numerical values will dominate. One way of weighing is to normalize the data residuals by the observed data. Many different weighing schemes can and have been developed for different applications.

3.5 Methods of Constraining the Solution

As discussed in an earlier section, the geophysical inverse problem is inherently ill-posed resulting in nonunique estimates of Earth model parameters. The method known as the *Backus-Gilbert approach* seeks average models to represent the solution of this problem.

Often the large number of unknowns can be approximated by a small number of coarse grid points (e.g., the Earth is assumed to be made up of a small number of discrete layers), which imposes well-posedness on the inverse problem. This constraint may be effected independently of the data **d**, and is called prior (or *a priori*) information or assumptions. Prior information can take many different forms. We describe some of them below.

3.5.1 Positivity Constraint. Based on the physics of the problem, our decision may be to constrain the model parameters so that they have no negative values. For example, seismic wave velocities and densities are always positive numbers, and the inversion results may be constrained to have only positive values for these parameters in seismic waveform inversion.

3.5.2 Prior Model. In many instances, we may have a very good reference (or prior) model. Thus, in the inversion, in addition to minimizing the misfit between observed and synthetic data, we may impose the restriction that the model not deviate significantly from the prior model $\mathbf{m}_p$. We can do this by modifying the error function $\mathbf{E}(\mathbf{m})$ to the following form:

$$E_3(\mathbf{m}) = (\mathbf{d}_{obs} - \mathbf{Gm})^T(\mathbf{d}_{obs} - \mathbf{Gm}) + \varepsilon(\mathbf{m} - \mathbf{m}_p)^T(\mathbf{m} - \mathbf{m}_p), \qquad \dots\dots\dots (3.27)$$

where ε is a weight (sometimes called *regularization weight*) that determines the relative importance of the model and the data error.

In general, this weight can be different for different model parameters. In some situations, we may have very good information on some model parameters, and we do not want them to vary significantly, while other model parameters may not be so well known, and must be allowed to deviate from $\mathbf{m}_p$. A convenient way of accomplishing this weighting is by employing a model covariance matrix, $\mathbf{C}_m$ (assuming that the prior probability density function—PDF—of the model is Gaussian), which gives a quantitative estimate of the certainty we have on the prior model. Thus, the error function $E_1(\mathbf{m})$ used in the maximum likelihood method can be modified to the following:

$$E_4(\mathbf{m}) = (\mathbf{d}_{obs} - \mathbf{Gm})^T\mathbf{C}_d^{-1}(\mathbf{d}_{obs} - \mathbf{Gm}) + (\mathbf{m} - \mathbf{m}_p)^T\mathbf{C}_m^{-1}(\mathbf{m} - \mathbf{m}_p). \qquad \dots\dots (3.28)$$

3.5.3 Model Smoothness. In many situations, we may require that the model be smooth (i.e., slowly varying in some direction). The smoothness of the model can be imposed by means of a model-weighting matrix $\mathbf{W}_m$ (which may have different forms based on the degree of smoothness desired), and the error function $E_2(\mathbf{m})$ can be modified to the following form:

$$E_5(\mathbf{m}) = (\mathbf{d}_{obs} - \mathbf{Gm})^T\mathbf{W}_d^{-1}(\mathbf{d}_{obs} - \mathbf{Gm}) + (\mathbf{m} - \mathbf{m}_p)^T\mathbf{W}_m^{-1}(\mathbf{m} - \mathbf{m}_p). \qquad \dots (3.29)$$

For all the error functions (E_1 through E_5), the formula for least squares solutions can be easily derived. Menke (1984) gives expressions for most of these.

3.6 Optimization Methods for Nonlinear Problems

As described in an earlier section, the principal objective in a nonlinear inverse problem is to locate a model (or models) for which a suitable defined objective function has a minimum. More precisely, we search for an optimal model that explains the observation reasonably well within the limits of noise in the data. The formal procedure for doing this is called *optimization* (e.g., Gill *et al.* 1981). The solution of an optimization problem is a set of allowed values of the variables for which an objective function assumes an "optimal" value. Optimization usually involves maximizing or minimizing; for example, we may wish to maximize profit or minimize weight. There are several methods of optimization, and the choice among them is based on the nature of the objective function and the type of constraints used. In a nonlinear inverse problem, the objective function is expected to have multiple minima of varying heights. For example, a hypothetical objective function of a single model parameter is shown in **Fig. 3.4**. The function is characterized by several minima with varying character. A point **m*** is a strong local minimum if $E(\mathbf{m}^*)$ is smaller than $E(\mathbf{m})$ for all points in its close neighborhood. Similarly, a point **m*** is a weak local minimum if $E(\mathbf{m}^*)$ is smaller than or equal to $E(\mathbf{m})$ for all points in its close neighborhood. The absolute minimum of $E(\mathbf{m})$ corresponds to the so-called *global minimum* of the objective function. Two classes of optimization methods can be identified. *Local optimization* methods search for a local minimum in the vicinity of a starting solution or a trial solution using local properties such as the first or second derivative of the objective function. *Global optimization* methods are generally stochastic algorithms that attempt to reach the so-called

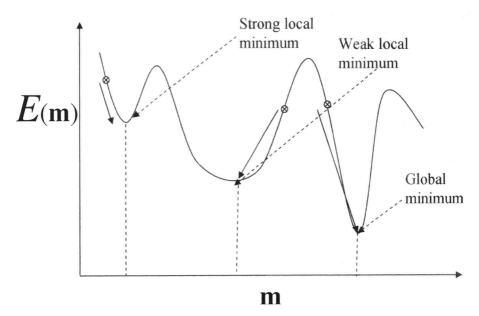

Fig. 3.4—A hypothetical error function for a single model parameter showing strong and weak local minima and a global minimum.

global minimum. Mathematicians and different application groups have proposed several variants of local and global optimization methods. We will describe below some of the local and global optimization methods that are commonly employed in seismic inversion.

3.6.1 Local Optimization Methods. The flow chart of a typical local optimization scheme is shown in **Fig. 3.5.** Most local optimization schemes are iterative algorithms, and the principal goal of all these algorithms is to ensure at each iteration that we attain a reduction in the objective function. In other words, these algorithms always attempt to travel in the downhill direction, and they are therefore referred to as *greedy algorithms*. It is fairly obvious that the local algorithms search for a local minimum in the neighborhood of the starting solution. Therefore, the choice of a starting solution is of paramount importance; a poor choice of the starting solution will cause the algorithm to get trapped in a local minimum.

Given a starting solution, the algorithm computes a search direction using a local property of the objective function, which determines an update or increment to the current

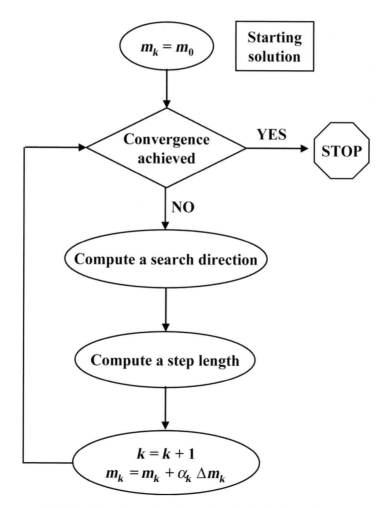

Fig. 3.5—A flow chart describing a local optimization scheme.

model. However, only a fraction of the computed update is applied, and a step-length factor is employed to ensure that the new objective function value is always smaller than the current value. Thus, the critical factors in our simple example algorithm (Fig. 3.5) are the calculations of the update and the step length. Optimization methods vary depending on the methods for computing these two parameters. The two most common algorithms are the steepest-descent and conjugate-gradient algorithms, which we describe below.

Steepest-Descent Algorithm. The steps for a steepest-descent algorithm are as follows:

1. Let $\mathbf{m}_k$ be the current model, and we want to choose a downhill direction $\Delta\mathbf{m}_k$ and step length α_k such that $E(\mathbf{m}_k + \alpha_k\Delta\mathbf{m}_k) < E(\mathbf{m}_k)$.
2. Using a Taylor series expansion we have $E(\mathbf{m}_k + \Delta\mathbf{m}_k) \approx E(\mathbf{m}_k) + (\nabla E(\mathbf{m}_k))^T\Delta\mathbf{m}_k$.
3. In a steepest-descent algorithm we choose $\Delta\mathbf{m}_k \approx -(\nabla E(\mathbf{m}_k))$.
4. Choice of α_k: Given a $\Delta\mathbf{m}_k$, we evaluate the objective function values for several trial values of α_k and choose the one that gives the minimum value of the objective function. Formally, this is done using a line-minimization algorithm.

This algorithm is fairly simple in principle and easy to implement as long as we can design an efficient scheme for computing partial derivatives of the error function with respect to the model parameters. Note, however, that the algorithm zigzags its way to the minimum: each successive direction is perpendicular to the previous one. Therefore, the algorithm becomes extremely slow when the objective function has narrow valleys (e.g., Press *et al.* 1992). This is because a steepest-descent algorithm computes very small step sizes in each iteration in a narrow valley.

Conjugate-Gradient Algorithm. The conjugate-gradient algorithm offers an advantage over a standard steepest-descent algorithm in that it avoids the slowdown caused by narrow valleys of the objective function. This is accomplished by computing updates along the so-called conjugate directions rather than along the direction of steepest descent (e.g., Gill *et al.* 1981; Tarantola 1987). The steps for a conjugate-gradient algorithm are as follows:

1. Compute synthetic data $\mathbf{d}_n = g(\mathbf{m}_n)$ and the derivative of data with respect to the current model given by $\mathbf{G}_n(\mathbf{m}_n)$.
2. Compute data residual $\Delta\mathbf{d} = \mathbf{d}_n - \mathbf{d}_{\text{obs}}$, and model residual $\Delta\mathbf{m}_n = \mathbf{m}_n - \mathbf{m}_{\text{pr}}$, where $\mathbf{m}_{\text{pr}}$ is the *a priori* model vector.
3. Compute the regularized objective function

$$E(\mathbf{m}_n) = \frac{1}{2}\left[\Delta\mathbf{d}_n^T\Delta\mathbf{d}_n + \Delta\mathbf{m}_n^T\Delta\mathbf{m}_n\right].$$

4. Compute the direction of steepest ascent $\gamma_n = [\mathbf{G}_n^T\Delta\mathbf{d}_n + \Delta\mathbf{m}_n]$.
5. Compute the conjugate direction $\varphi_n = \gamma_n + \sigma_n\varphi_{n-1}$, such that $\varphi_0 = \gamma_0$ and

$$\sigma_n = \frac{(\gamma_n - \gamma_{n-1})^T\gamma_n}{\gamma_{n-1}^T\gamma_{n-1}}.$$

6. Compute optimum step length μ_n using a linear search.
7. Update model $\mathbf{m}_{n+1} = \mathbf{m}_n - \mu_n\varphi_n$.
8. Go to Step 1.

3.6.2 Global Optimization Methods. Recently (because of the advent of powerful and relatively inexpensive computers), global optimization methods have been applied to several geophysical problems. Unlike local optimization methods, these methods attempt to find the global minimum of the misfit function. Most of the global optimization algorithms are stochastic in nature and use more global information about the misfit surface to update their current position. The convergence of these methods to the globally optimal solution is not guaranteed for all the algorithms. Only for some of the algorithms—and under certain conditions—is convergence to the globally optimal solution statistically guaranteed. Also, with real observational data, it is never possible to know whether the derived solution corresponds to the global minimum or not. However, our experience indicates that we are able to find many good solutions starting with only poor initial models using global optimization methods (Sen and Stoffa 1995). Most global optimization methods are less greedy than the well-known local optimization methods, in that during iterative optimization worse solutions are occasionally accepted, which allow these algorithms to avoid local minima. There are several variants of global optimization methods; detailed descriptions of some of these approaches as applied to geophysical inversion can be found in Sen and Stoffa (1995). Here we provide brief descriptions of two commonly used global optimization methods, namely, *simulated annealing* (SA) and the *genetic algorithm* (GA).

Simulated Annealing. SA was first proposed by Kirkpatrick *et al.* (1983). It is analogous to the natural process of crystal annealing, in which a liquid gradually cools to a solid state. The SA technique starts with an initial model m_o, with associated error or energy $E(m_o)$. It draws a new model $E(m_o)$ from a flat distribution of models within the predefined limits. Note that each model parameter can be bounded by different limits. The associated energy (objective function value) $E(m_{new})$ is then computed, and compared with $E(m_o)$. If the energy of the new state is less than the initial state, the new state is considered to be good. In this case, the new model is accepted, and it replaces the initial model unconditionally. However, if the energy of the new state is larger than the initial state, $E(m_{new})$ is accepted with the probability of $\exp\{-[E(m_{new}) - E(m_o)]/T\}$, where T is a control parameter called annealing temperature that controls whether the "bad" model should be carried over to the new model. This completes one iteration. It is the rule of accepting with a probability that makes it possible for SA to be able to jump out of the local minima. The same process is repeated for a large number of times, with the annealing temperature gradually decreasing according to a predefined scheme. The hope is that with a carefully defined cooling schedule, a global minimum can be found. One may choose a linear or a logarithmically decreasing cooling scheme. The tradeoff here is between the computation cost and the accuracy of the result. Fast cooling will fail to produce a crystal (the algorithm gets stuck in a local minimum), while slow cooling takes a long time but may eventually find the global minimum.

To speed up the annealing process without much sacrifice in the solution, a variant of SA, called *very fast simulated annealing* (VFSA) was proposed by Ingber (1993). VFSA differs from SA in a number of ways. The new model is drawn from a temperature-dependent Cauchy-like distribution centered on the current model **(Fig. 3.6)**. This change has two fundamental effects. First, it allows for larger sampling of the model space at the early stages of the inversion when the temperature is high and much narrower sampling in the model space as the inversion converges when the temperature decreases. Second, each model parameter can have its own cooling schedule and model space-sampling scheme. This, therefore, allows for individual control for each parameter and the incorporation of *a priori* in-

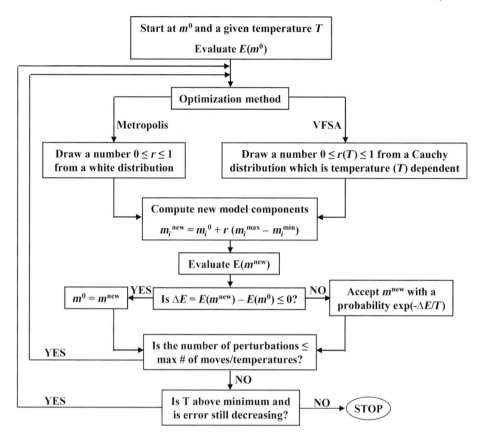

Fig. 3.6—A flow chart of a simulated annealing algorithm—note that the VFSA draws models from a Cauchy-like distribution, the shape of which changes with iteration.

formation. For many geophysical inversion applications, VFSA has been demonstrated to have excellent performance (Sen and Stoffa 1995).

Genetic Algorithm. GAs are so named because they emulate the biological processes of evolution and are based on the principle of survival of the fittest. In a GA, the model parameters are generally coded in a binary form (**Fig. 3.7a**). The algorithm starts with a randomly chosen population of models called *chromosomes*. The second step is to evaluate the fitness values of these models. Note that unlike our preceding discussion of optimization, in which we attempt to minimize an objective function, a GA searches for the maximum of a suitably defined fitness function that measures similarities between the observed and synthetic data. After the selection process, the three genetic processes of selection, crossover, and mutation are performed upon the models in sequence.

In *selection*, models are copied in proportion to their fitness values based on a probability of selection. Thus, during this process, models with higher fitness values (lower error values) have a higher probability of getting selected.

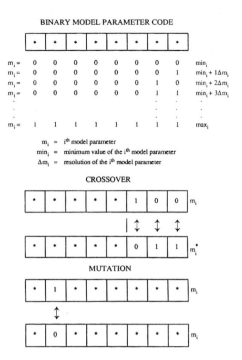

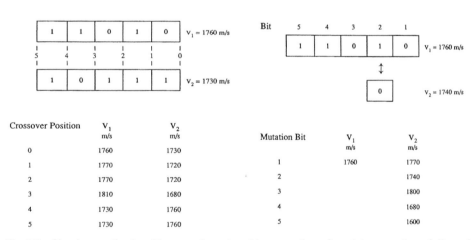

Fig. 3.7—Simple genetic algorithms are based on binary coding of model parameters, followed by the processes of selection, crossover, and mutation.

Crossover acts on the selected pairs of models. This operation picks a crossover site within the paired strings and exchanges the bits between the two models to the right of the crossover site on the basis of crossover probability (Fig. 3.7b). Thus, the crossover process results in two new models (or "children").

Mutation, which involves changing a bit at random based on a mutation probability, is applied to the models to maintain diversity.

After execution of these three processes, the new models are compared to the previous generation and accepted on the basis of an update probability. The procedure is repeated until convergence is reached (i.e., when the fitness of every model becomes close to the fitness of every other model).

Note that the repeated applications of selection, crossover, mutation, and update do not necessarily guarantee convergence to the global maximum of the fitness function. Nonetheless, the algorithm has been found to be successful in many optimization problems, including those from geophysics.

Hybrid optimization methods that take advantage of both local and global optimization methods have also been proposed. Some strategies for hybrid optimization are discussed in Chunduru *et al.* (1997).

Chapter 4

Inversion of Post-Stack Data: Wavelet and Acoustic Impedance

For many seismic processing applications, it becomes necessary to derive an estimate of the seismic wavelet. Because the character of the wavelet is imprinted on seismic traces, it is important to understand its shape in order to decipher the properties of the Earth's interior from seismic traces. In spite of the fact that the wavelet is time-varying and is expected to be spatially varying, an overall knowledge of the wavelet is crucial to enhancing resolution for better imaging of structures and predicting lithology and fluid content. The most common practice is to invert post-stack seismic data for wavelets. As described in Chapter 2, a post-stack trace emulates a zero-offset or normal-incidence seismogram, which can be simulated using a convolution model (Eq. 2.35) assuming a locally 1D Earth model. Most seismic data contain noise; therefore, Eq. 2.35 is modified to the following form:

$$S_w(t) = W(t) \otimes \sum_{j=1}^{N} R_{pp}^j \delta(t - t_j) + n(t), \qquad \ldots\ldots\ldots\ldots\ldots\ldots\ldots\ldots \text{(4.1)}$$

where $W(t)$ is the wavelet (generally a smooth function of time), R_{pp}^j is the P-wave normal-incidence reflection coefficient at interface j, t_j is the two-way normal reflection time from interface j, N is the total number of layers, $n(t)$ is the noise as a function of time, and the symbol $\otimes$ represents a convolution operation. Also recall that the normal-incidence reflection coefficient is given by

$$R_{pp}^i = \frac{Z_{i+1} - Z_i}{Z_{i+1} + Z_i}, \qquad \ldots\ldots\ldots\ldots\ldots\ldots\ldots\ldots\ldots\ldots\ldots\ldots\ldots\ldots \text{(4.2)}$$

where $Z_i = \alpha_i \rho_i$ is the acoustic impedance. In the frequency domain, the convolution operation is replaced by a multiplication. Given a seismic trace, three inverse problems are identified:

1. Estimation of the wavelet when the reflection coefficient is known.
2. Estimation of reflection coefficients or acoustic impedances when the wavelet is known.
3. Simultaneous inversion for acoustic impedance and wavelet.

For the general case of simultaneous inversion of impedance and wavelet, we are faced with the problem of estimating two multipliers from a given product in the presence of noise. Given an Earth model, the source wavelet can be estimated from a normal-incidence seismic trace using principles of inverse theory described in Chapter 3. Even if the wavelet is known, however, the estimation of acoustic impedance is not trivial. Algorithms do exist for the simultaneous estimation of wavelet and acoustic impedance, and these have been found to be effective in many situations. We will discuss these techniques together with the inherent nonuniqueness issues related to the estimation problem.

4.1 Nonuniqueness and Deconvolution

An example of the convolution process for constructing a seismogram using a wavelet and an impedance function is shown in **Fig. 4.1**. Note that a wavelet is a smoothly varying function, while the reflectivity (computed from the impedance profile) is a series of delta functions placed at the two-way normal time of each reflector. The spectra of the wavelet and

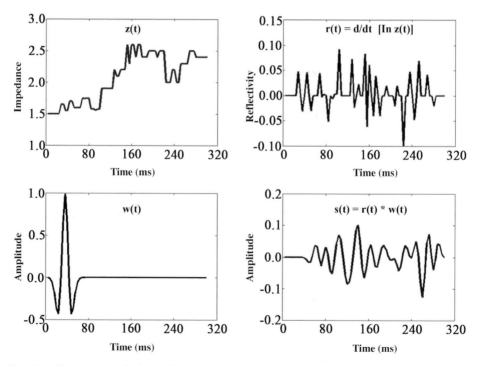

Fig. 4.1—Given an impedance profile and a source wavelet, a post-stack or zero-offset synthetic seismogram is computed by a convolution. The reflection coefficient series is computed from the impedance profile, which is convolved with the wavelet (lower left panel) to generate a seismogram (lower right panel).

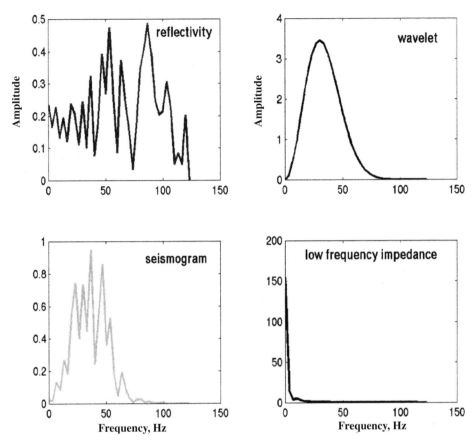

Fig. 4.2—The inverse problem of impedance inversion is underdetermined. The upper panel displays the spectrum of the reflectivity (left) and the wavelet (right). The lower panel shows the reflectivity spectrum (left) after the convolution process. A low-frequency trend is provided as *a priori* information in an impedance inversion (lower right panel).

the reflectivity series for a synthetic example are shown in **Fig. 4.2**. We observe that the wavelet is band-limited, while the reflectivity series is broadband. Because the convolution is equivalent to multiplication in the frequency domain, the spectrum of the resulting seismogram is band-limited as well (i.e., the seismogram lacks frequencies in the low- and high-frequency ranges because of filtering caused by the wavelet). We can imagine the complexity of the problem further when we take into account the loss of high frequencies of the wavelet caused by attenuation. In other words, the time series cannot be assumed to be stationary. Even under the stationarity assumption, the data do not contain all the frequencies; therefore, we must expect to derive a band-limited reflectivity estimate unless prior constraints are imposed. One popular approach is to set up the optimization problem such that reflectivity is constrained to be spike-like and the resulting reflectivity series is sparse. These constraints can be imposed through the model norm in the definition of the objective function (Eqs. 3.27 and 3.28).

Given a source wavelet, the process of removing the wavelet is called *deconvolution*—a process commonly applied to seismic data to generate high-resolution images. In principle, deconvolution can be achieved by spectral division, as given by

$$S_\delta(\omega) = \frac{S_w(\omega)}{W(\omega)}, \quad \dots\dots\dots\dots\dots\dots\dots\dots\dots\dots\dots\dots\dots\dots\dots\dots (4.3)$$

where $S_\delta(\omega)$ is the deconvolved seismic trace, $S_w(\omega)$ is the spectrum of the data, and $W(\omega)$ is the spectrum of the wavelet. Obviously, the division process described in Eq. 4.3 is highly unstable because of the zeros in the spectrum of the source wavelet. To avoid this problem, a common practice is to add a small nonzero value in the denominator, resulting in the following equation:

$$S_\delta(\omega) = \frac{S_w(\omega)}{W(\omega) + \varepsilon}, \quad \dots\dots\dots\dots\dots\dots\dots\dots\dots\dots\dots\dots\dots\dots\dots (4.4)$$

where ε is a small number.

A more stable algorithm, proposed by Wiggins (1977), cast the deconvolution problem as a maximization of the following objective function:

$$\nu = \frac{\sum\limits_{i-1}^{n} (S_w(i\Delta t))^4}{\left(\sum\limits_{i-1}^{n} (S_w(i\Delta t))^2\right)^2}, \quad \dots\dots\dots\dots\dots\dots\dots\dots\dots\dots\dots\dots\dots (4.5)$$

where n is the number of time samples and Δt is the sampling interval. The norm defined in Eq. 4.5 is called the *varimax*. Maximizing the varimax norm is equivalent to minimization of *entropy*, and the process is popularly known as *minimum entropy deconvolution, or MED*. In Eq. 4.5, the sums are simply the scaled moments of the seismic amplitudes, and because only the even moments are used, the varimax norm is insensitive to the polarity or sign of the seismic amplitude. The norm has high values for distribution with a large fourth moment, where more samples have high amplitudes. However, the square of the second moment reduces the value of the norm for increasing numbers of peaks. Therefore, the norm yields higher values, as the distribution is concentrated into a few high-amplitude peaks, achieving both spikiness and simplicity. The varimax norm is maximum for data consisting of all zeros except for a single spike.

The minimization of the entropy (given a source wavelet) can be carried out using linear or nonlinear inversion schemes, as described in Chapter 3.

4.2 Wavelet Estimation

The basic assumption in wavelet estimation is that an estimate of the reflectivity is known and attempts are made to estimate a smooth wavelet. Oldenburg *et al.* (1981) outlined detailed algorithms for achieving this by using linear inverse theory. In one of the approaches, they represented the wavelet $W(t)$ in terms of the basis function

$$W(t) = \sum_{i=1}^{N} \alpha_i \psi_i(t), \quad \dots\dots\dots\dots\dots\dots\dots\dots\dots\dots\dots\dots\dots\dots (4.6)$$

where α_i are the coefficients. Further, the α_i values corresponding to small eigenvalues (corresponding to high-frequency oscillations) were discarded to ensure smoothness. One additional constraint was to ensure that the squares of the second and fourth derivatives of the wavelet are minimum, which ensures the flattest, smoothest models of the wavelet. Oldenburg et al. (1981) employed three different schemes to solve the problem—cast as a linear inverse problem—and successfully applied the algorithm to field seismic data.

Recently, Wood (1999) designed an algorithm for joint inversion of reflectivity and the wavelet in the frequency domain. His algorithm assumes that the amplitude spectrum of the wavelet is known as given by the amplitude spectrum of the seismic trace and solves for the phase spectrum of the wavelet. Rather than using a varimax or entropy criterion, the algorithm uses the following norm:

$$L_\alpha[W(t)] = \frac{1}{S_{max}} \left(\sum_{i=1}^{n} (W(i\Delta t))^\alpha \right)^{1/\alpha}, \qquad \dots\dots\dots\dots\dots\dots\dots (4.7a)$$

where α can take any integer value, S_{max} is the maximum of the distribution, and only the absolute values of the amplitude are considered in Eq. 4.7. The maximum value of all these norms is unity. The algorithm inverts for the wavelet-phase components and band-limited reflectivity such that the wavelet convolved with the reflectivity results in a seismogram with the lowest L_1 norm. The nonlinear optimization problem was solved using VFSA described in Chapter 3.

The most common approach to deriving wavelets is based on well-log data that provide the "true" reflectivity series. In other words, given the compressional wave velocity and density logs, we compute acoustic-impedance logs, which are mapped into a normal-incidence reflection coefficient series (Eq. 4.2). An initial guess of the wavelet is convolved with the reflectivity series, and a synthetic normal-incidence trace is generated. The difference between the observed and the synthetic trace is minimized using a suitably chosen norm with a smoothness constraint. The algorithms described in Chapter 3 are used to find an optimal solution. A field data example of wavelet estimation is shown in **Fig. 4.3**. A commercial package (Jason Workbench) was used to derive the wavelet; the stack section (second from left) was derived from data collected over offshore Oregon. A well-log-derived acoustic-impedance profile (right-most column in Fig. 4.3) was used in the inversion. The derived wavelet is shown in the left panel of Fig. 4.3.

4.3 Impedance Estimation

Initial attempts at estimating impedances were based on deriving reflection coefficient series that were based on a linear inversion scheme with spikiness constraints. The acoustic impedances were then derived from the reflection coefficients. Recall that the normal-incidence plane-wave reflection coefficient at a layer boundary k denoted by R_{pp}^k is given by Eq. 4.20, such that the acoustic impedance $Z_k = \alpha_k \rho_k$. Therefore,

$$Z_{k+1} = Z_k \left(\frac{1 + R_{pp}^k}{1 - R_{pp}^k} \right) = Z_1 \prod_{j=1}^{k} \left(\frac{1 + R_{pp}^j}{1 - R_{pp}^j} \right). \qquad \dots\dots\dots\dots\dots\dots (4.7b)$$

Given knowledge of the impedance of the shallowest layer, the impedances of the successively deeper layers can be derived using the recursion formula given in Eq. 4.7b.

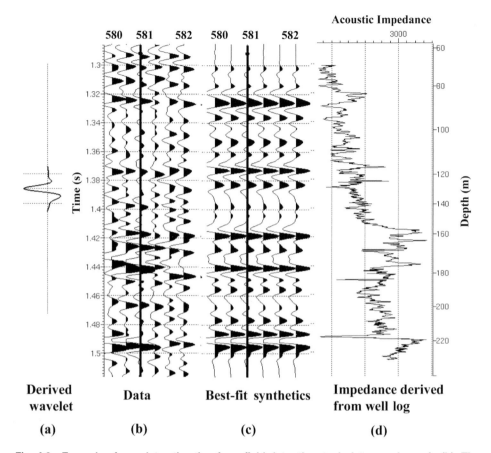

Fig. 4.3—Example of wavelet estimation from field data: the stack data are shown in (b). The well-log-derived impedance profile (d) is used in the estimation. The estimated wavelet is shown in (a), while the synthetic seismograms are shown in (c).

The difficulties with using a recursion scheme are well known (Cooke and Schneider 1983), in that the following conditions must be met: The source wavelet must be completely removed; noise must be absent; all multiple reflections must be removed; and the spherical spreading and transmission losses, including out-of-plane reflections, must be removed. Unfortunately, such conditions are rarely met; therefore, a model-based inversion of seismic traces for impedance estimation is often preferred. Before we discuss the model-based inversion algorithms, we must revisit the nonuniqueness issue, which we introduced earlier in the section on wavelet estimation.

We noted earlier that because of the band-limited nature of the source wavelet, a typical seismogram is devoid of low- and high-frequency information. Therefore, the inversion of band-limited response can only be expected to yield an approximate acoustic-impedance log. The band limitation introduces an undesirable effect, namely, the violation of causality. It can be shown analytically (Ghosh 2000) that

- The upper frequency limit imposes a seismic resolution restriction on the acoustic impedance.
- The lower limit imposes a restriction such that inversion cannot generate absolute values, only relative values.

Fig. 4.4 shows expected results from the impedance inversion of data devoid of low frequency alone (upper panel) and of data with missing low and high frequencies. For a single interface problem, we can clearly observe that an impedance inversion of seismic traces can only resolve band-limited impedance contrasts. The low-frequency component needs to be incorporated *a priori* in the inversion algorithm. Only then can we expect to derive absolute impedances. Low-frequency components are generally provided from the NMO velocity analysis, well-log-information, or regional trends. Prior constraints also need to be imposed to ascertain sparsity and spikiness in the reflectivity series or step-like features in the impedance profiles. Such algorithms are generally referred to as *sparse-spike inversion*.

A typical flow chart for a model-based impedance inversion algorithm is shown in **Fig. 4.5**. The forward problem in this application is a simple convolution model. Both the local (conjugate gradient and other variants) and global optimization (simulated annealing and genetic) algorithms have been successfully employed in the impedance estimation problem. The crucial part of developing an optimization approach is the choice of the objective function, which may take the following form for the impedance estimation problem:

$$E(\mathbf{m}) = E_d + cE_m, \quad \ldots\ldots\ldots\ldots\ldots\ldots\ldots\ldots\ldots\ldots\ldots\ldots\ldots\ldots (4.8)$$

where E_d is the data norm, E_m is the model norm, and c is the regularization weight. Generally, an L_2 norm for data misfit is used. The model norm, however, contains several parts. They include a part that minimizes the differences between the prior and the current model, incorporating a low-frequency trend (generally an L_2 norm), an L_1 norm for the reflectivity, ensuring spikiness, and possibly a model flatness term. The regularization weight is chosen by trial and error or by an analytic formula applicable for a particular optimization algorithm. Local optimization methods require evaluation of a sensitivity matrix, but global optimization methods are generally derivative-free. An example of inversion of the noise-free synthetic data for a model shown in Fig. 4.1 is displayed in **Fig. 4.6**. The seismogram was inverted using a conjugate gradient algorithm. Given a reasonably low-frequency trend, we were able to derive a good estimate of the impedance profile, resulting in excellent data fit.

In our synthetic example shown in Fig. 4.6, we were able to obtain a realistic low-frequency trend, because the true model was known. For application to real data, the most common practice is to make use of well-log information to develop background models for impedance. Because the well-log data are available at sparse locations, they are interpolated at all other CMP locations. The interpolation, however, is not trivial in areas with laterally varying geology. It requires interpreter's skills to mark horizons along which well logs are interpolated. A flow diagram for impedance inversion is displayed in **Fig. 4.7**. The steps are outlined below:

- Estimate a wavelet using well log and stacked seismogram from well location— wavelets are derived from multiple well locations, and an average wavelet is generally used in the impedance inversion.

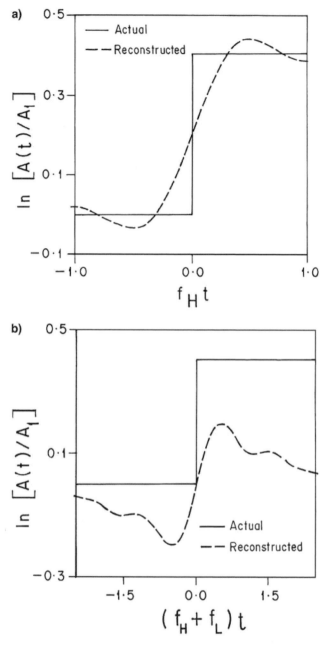

Fig. 4.4—Plots of normalized travel time vs an impedance discontinuity involving a jump from A1 to A2 (=1.5 A1), along with its band-limited reconstruction, both expressed in terms of logarithm of the impedance scaled with respect to A1. (a) A low-pass seismic wavelet with a uniform spectrum between –fH and fH; fHt denotes the normalized travel time. (b) A band-pass wavelet with a uniform spectrum between fL (10 Hz) and fH (100 Hz) and between –fL and –fH; (fl + fH)t is the normalized travel time (from Ghosh 2000; reprinted with permission from the Soc. of Exploration Geophysicists).

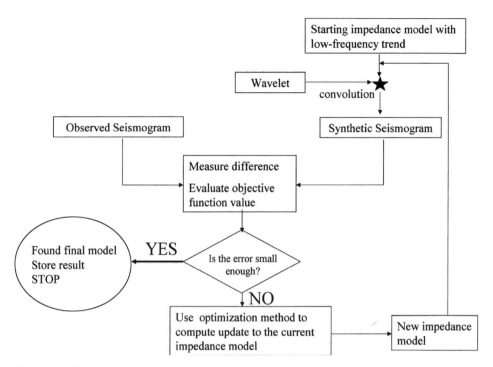

Fig. 4.5—A flow chart describing an impedance estimation scheme based on model-based inversion.

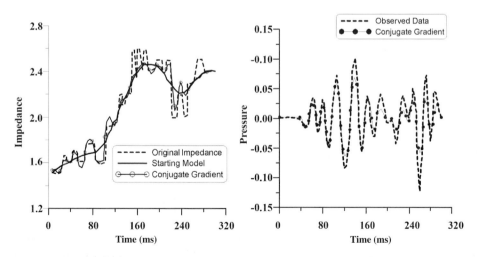

Fig. 4.6—Results from impedance inversion of a synthetic trace shown in 4.1a: the low-frequency trend is shown by the solid line. Inversion was carried out with a conjugate gradient scheme. Note that the impedance estimates are close to the true values of the impedance (left panel) and the data fit is excellent (right panel).

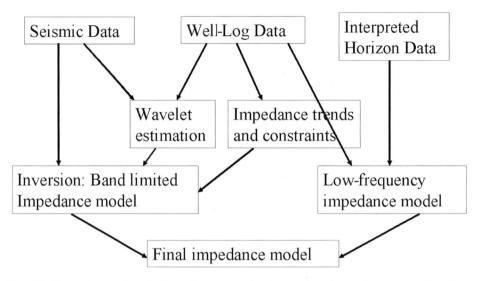

Fig. 4.7—General procedure for impedance inversion by combined use of seismic and well-log data.

- Determine impedance trends and constraints from well logs, and using the inverted wavelet, derive estimates of band-limited impedance (relative impedance) at all CMP locations.
- Interpret horizons using well logs and the stack section and derive a low-frequency trend that is laterally consistent. Merge the low-frequency impedance model with the inverted band-limited impedance model to estimate the final impedance model.

An example of post-stack inversion of a 2D seismic line from Hydrate Ridge offshore Oregon is shown in **Fig. 4.8**. The shallow sediments in the Hydrate Ridge area are characterized by the occurrence of gas hydrates*; the base of the gas hydrate is often conspicuous in a seismic section as a very bright reflection called the *bottom simulating reflection* (BSR), which roughly parallels the topography of the seafloor. The BSR essentially marks the boundary between gas hydrates (frozen gas) and free gas. The U. of Texas Inst. for Geophysics and Oregon State U. conducted a 3D seismic survey in the Hydrate Ridge region; several wells were also drilled in this area by the Ocean Drilling Program (ODP). A 2D seismic line together with four wells is shown in Fig. 4.8a. The area is structurally complex, and therefore it is not straightforward to interpret horizons as required by the impedance inversion algorithm. The source wavelet was derived at one of the well locations (Fig. 4.3) using an inversion algorithm described in an earlier section. The results from relative acoustic-impedance inversion are shown in Fig. 4.8b; the base of the hydrate is characterized by relatively low impedance. Another anomalous region—characterized by low impedance—is

---

* Gas hydrates are ice-like crystalline structures of a water lattice with cavities, which contain gas molecules.

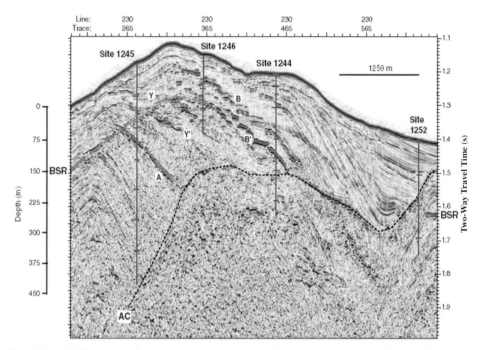

Fig. 4.8a—A 2D seismic line (stack section) collected over hydrate ridge offshore Oregon. It shows an east/west vertical slice through the 3D seismic data showing the stratigraphic and structural setting of Sites 1244, 1245, 1246, and 1252. A bright reflection paralleling the topography of the seafloor marks the base of gas hydrate (BSR); free gas is expected below the BSR (Shipboard Scientific Party, Leg 204, 2002).

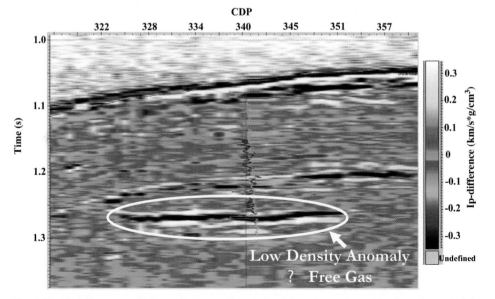

Fig. 4.8b—Relative acoustic-impedance section derived from inversion of the line shown in (a); the wavelet shown in Fig. 4.3 was used in the inversion. The P-wave velocity and density logs are superimposed on the section. An anomalous low-impedance zone below the BSR is marked with an ellipse.

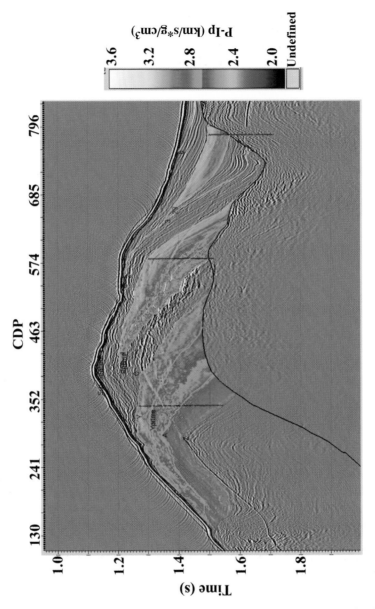

Fig. 4.8c—Final acoustic impedance estimates: Horizons were interpreted in the seismic section together with the well logs. Note that this region is structurally complex, and we only have four well logs; therefore, it was not possible to extend the horizons everywhere in the section. Well logs were then interpolated along the horizons. The final impedance section obtained after merging the relative impedance section with the low-frequency background derived from well logs is shown here.

marked in the figure; this region is also characterized by the accumulation of free gas. Fig. 4.8c shows interpreted horizons between the wells. Within these horizons, the well-log-derived impedances were interpolated, and low-frequency impedance estimates were made. Final estimates of acoustic impedance within the interpreted horizons are displayed as color sections in Fig. 4.8c.

4.4 Estimation of Petrophysical Parameters—Interpretation of Results

The ultimate goal of seismic inversion is to obtain a map of porosity, lithology, and other reservoir bulk properties for use in reservoir model-building. The impedance volumes derived from seismic inversion are transformed into porosity maps using transformations called *trends* (Dvorkin and Alkhater 2004), which are built from borehole measurements that establish a relationship between the two parameters (impedance and porosity). In other words, an empirical porosity trend is determined from sonic, density, and porosity curves, which can be applied to acoustic-impedance volume derived from seismic inversion for mapping porosity in 3D. One important goal of this transformation is to account for differing resolutions in seismics and well logs using a procedure known as *upscaling*. As described in Chapter 1, the seismic wavelength is usually much larger than the scale of variation in rock properties measured in well-log data. One approach to achieving the goal is to conduct upscaling using the quarter wavelength as an averaging window (Dvorkin and Alkhater 2004), such that a running mean filter is applied to the porosity data while effective impedance is calculated using Backus (1962) averaging.

As an example, we describe an excellent case study presented by Dvorkin and Alkhater (2004) using a 3D seismic volume and well logs from the Troll field. **Fig. 4.9a** shows a 2D cross section bounded by two wells within a time window of 500 ms (1400 ms to 1900 ms). The reservoir is located between two seismic horizons, A and B, and the gas/oil contact is located at the flat reflector C. The trends derived from the two wells at the log scale and in the upscaled form are shown in Fig. 4.9b. The inverted impedance section and a porosity map derived from it using the trends shown in Fig. 4.9b are displayed in the upper and lower panels of **Fig. 4.10.** Dvorkin and Alkhater (2004) noted that the porosity section is qualitatively different from the impedance section, in that it does not reflect the sharp horizontal impedance contrast at the flat gas/oil contact reflector. This is because, before using the impedance-porosity transforms, Dvorkin and Alkhater (2004) identified the pore fluid and then applied the transforms selectively.

Although the transformation methods may vary, it is now a common practice in the industry to generate porosity maps from 3D impedance volumes obtained from post-stack seismic inversion results.

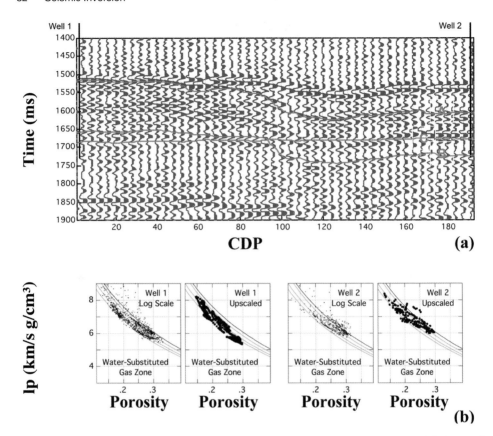

Fig. 4.9a—The seismic section (full stack) between two vertical wells. The reservoir is located between reflectors A and B. Reflector C marks the transition from gas to oil. (b) Impedance-porosity crossplots of water-substituted well-log data at the original log scale as well as up-scaled. The curves in the crossplots are from effective-medium models. The red curves are from the uncemented sand model with clay content of 0, 10, and 20%. The black curve is from the constant cement model. In the models, the critical porosity is 0.4, the coordination model is 8, and the cemented porosity is 0.375 (Dvorkin and Alkhater 2004. Reprinted with permission from the European Assn. of Geoscientists and Engineers and J. Dvorkin).

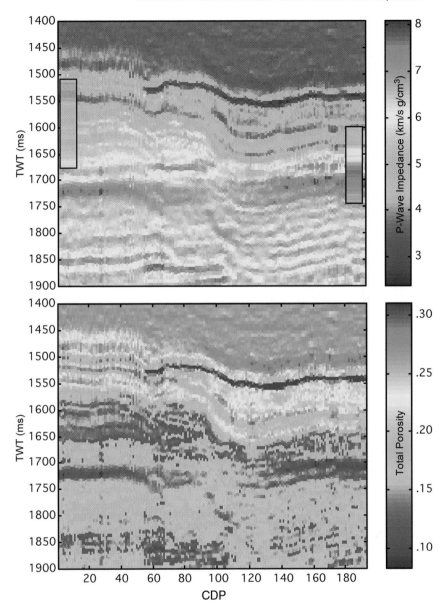

Fig. 4.10—Impedance inversion section (top) and predicted porosity section (bottom). The inserted strips in the upper frame mark the impedance in the wells (Dvorkin and Alkhater 2004. Reprinted with permission from the European Assn. of Geoscientists and Engineers and J. Dvorkin).

Chapter 5

Prestack Inversion

The amplitudes in normal-incidence seismograms, which are generally approximated by post-stack data, are sensitive to contrasts in acoustic impedance across layer boundaries. The post-stack inversion methods, such as the sparse-spike inversion described in Chapter 4, estimate acoustic impedances, which can often be related to porosity. Post-stack data, however, cannot resolve compressional wave velocity and density separately, and are not at all sensitive to shear-wave velocities. The expressions for angle-dependent reflection coefficients show that seismic wave amplitudes are sensitive to changes in shear-wave velocities at nonnormal angles of incidence. The goal of prestack inversion is to make use of reflection amplitude, traveltime and waveform data at nonnormal incidence to estimate acoustic impedance and Poisson's ratio so that a more robust interpretation of lithology and fluid content can be made.

Over the last three decades, the exploration industry has witnessed the development of a suite of methods that make use of prestack data for parameter estimation. They include amplitude variation with offset (AVO), elastic impedance, and full waveform inversion. In this chapter, we will describe all three approaches in detail.

5.1 Amplitude Variation With Offset

It has been a long-term goal of geophysics to predict lithology and estimate fluid content from seismic data. The occurrence of a bright spot or amplitude anomaly in a true amplitude-processed section such as that in Fig. 4.8 has been used extensively for this purpose. Such bright spot phenomena are often associated with the occurrence of gas in a sand-shale sequence. This is because introduction of gas to a porous sand unit generally decreases the unit's acoustic impedance relative to the surrounding lithologic units, which results in a high-reflectivity zone or bright spot. Although bright spots are often associated with gas sands, they may also be caused by the presence of lithology such as coal, over-pressure shales, and high-porosity sands. It is believed that such an ambiguity can be overcome by prestack analysis. As described in Chapter 1, the addition of gas into water-saturated sand reduces the P-wave velocity and density of the sand, but causes minimal change in the shear-wave velocity, which depends on the formation's shear modulus and density. Replacement of brine by gas does not affect the shear modulus significantly; thus, there is minimal change in shear-wave velocity. On the other hand, lithologic changes generally cause changes in P- as well as S-wave velocities. Additional information on elastic properties that is necessary for such discrimination can be obtained from prestack data only.

The most popular approach to prestack data analysis is the examination of amplitude variation with offset or angle (AVO) (Castagna and Backus 1993). It assumes that the variations in seismic wave amplitude with offset are related directly to plane-wave reflection coefficients. In Chapter 2, we outlined several factors affecting seismic wave amplitudes. Application of AVO analysis requires that the seismic data be processed while preserving true amplitude and correcting for amplitude variation caused by other factors such as spherical spreading. The amplitude variations are then modeled using a linearized reflection coefficient model given by Eq. 2.27 and 2.28. Note that AVO analysis makes use of a linearized model for easy interpretation of the results.

Motivations for using reflection coefficients in seismic interpretation were first documented in the classic work of Ostrander (1984), who first reported that AVO is affected strongly by the relative values of Poisson's ratio across an interface separating two rock layers. He demonstrated that high-porosity gas sands exhibit abnormally low Poisson's ratios, resulting in an increase in reflected P-wave amplitude with angle of incidence. This study formed the basis for classification of AVO types for a range of AVO effects associated with the gas sands normally encountered in exploration (Rutherford and Williams 1989). They include high-impedance sands, near-zero-impedance contrast sands, and low-impedance sands. AVO curves computed for a shale-sand interface with varying impedances of sand, using exact expressions for reflection coefficients (e.g., Eq. 2.25), are shown in **Fig. 5.1**. The top curve represents the AVO curve for sand with higher impedance than the shale layer. It shows that the reflection coefficient is positive at normal incidence and then decreases in amplitude with offset; it can also change polarity at large offset. Such an AVO is known as a Class 1 AVO. The two curves in the middle in Fig. 5.1 are computed for sand with nearly the same impedance as that of the shale layer. This represents Class 2 AVO, which may or may not correspond to amplitude anomalies on stacked data. Class 3 AVO, as shown in Fig. 5.1, is characterized by a large negative reflection coefficient at normal incidence, which increases gradually with increasing offset or angle. There are no polarity changes associated with class 3 AVO, and thus CMP stacking shows bright amplitude corresponding to such a reflection event. Note also that the AVO gradients are different for the three types of AVO. These characteristics are used in the interpretation of results from AVO inversion.

5.1.1 AVO Inversion. The forward model for AVO analysis is based on either an Aki and Richards approximation of the Zoeppritz equation given by

$$Rpp(p) \cong \frac{1}{2}(1 - 4\beta^2 p^2)\frac{\Delta\rho}{\rho} + \frac{1}{2\cos^2 i}\frac{\Delta\alpha}{\alpha} - 4\beta^2 p^2\frac{\Delta\beta}{\beta} \quad \dots\dots\dots\dots (5.1)$$

(terms in the above equation are given in Chapter 2, Eq. 2.27) or a two- or three-term approximation of the Zoeppritz equation at an incidence angle given by

$$R_{pp}(\theta) = R_0 + \left(A_0 R_0 + \frac{\Delta\sigma}{(1-\sigma)^2}\right)\sin^2\theta + \frac{1}{2}\frac{\Delta\alpha}{\alpha}(\tan^2\theta - \sin^2\theta), \quad \dots\dots (5.2)$$

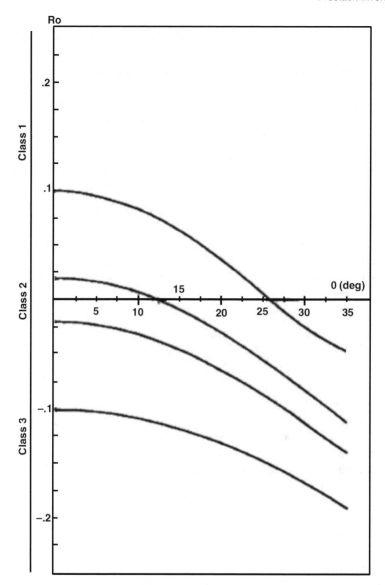

Fig. 5.1—Zoeppritz P-wave reflection coefficients for a shale-gas sand interface for a range of values of normal-incidence reflection coefficients. The Poisson's ratio and density of shale were assumed to be 0.38 and 2.8 g/cm³, respectively. The Poisson's ratio and density of gas sand were assumed to be 0.15 and 2.0 g/cm³, respectively. The three classes of AVO (class I, class II, and class III) are marked in the figure (from Rutherford and Williams 1989; reprinted with permission from the Soc. of Exploration Geophysicists).

with

$$A_0 = B_0 - 2(1 + B_0)\left(\frac{1 - 2\sigma}{1 - \sigma}\right)$$

$$B_0 = \frac{\Delta\alpha/\alpha}{\left(\dfrac{\Delta\alpha}{\alpha} + \dfrac{\Delta\rho}{\rho}\right)},$$

where R_0 is the normal-incidence reflection coefficient and σ is Poisson's ratio. Equation 5.2 is sometimes written as

$$R_{pp}(\theta) = A + B \sin^2 \theta + C \tan^2 \theta. \quad \dots\dots\dots\dots\dots\dots (5.3)$$

The coefficients A, B, and C are often referred to as the *AVO intercept, AVO gradient,* and *AVO curvature,* respectively.

An AVO inversion algorithm essentially derives, depending on the equation used, either the fractional changes in compression wave velocity, shear-wave velocity and density, or the AVO terms A, B, and C. There are two approaches to AVO inversion: one is based simply on fitting amplitudes at every time sample, and the other is essentially a waveform inversion that uses Eq. 5.1, 5.2, or 5.3 for computing seismograms.

5.1.2 AVO Inversion by Amplitude Fitting. Basic principles behind this algorithm are described in Chapter 3 (Eq. 3.17). Using either Eq. 5.1 or a two-term form of Eq. 5.2, one defines the following standard linear system:

$$\mathbf{d} = \mathbf{Gm},$$

where the data vector $\mathbf{d}$ contains reflection amplitude at different offsets or angles, and the model vector $\mathbf{m}$ contains either the fractional changes in material properties or the AVO intercept and gradient.

The input data consist of NMO-corrected seismograms in the offset, angle, or ray-parameter domain. At each time sample, amplitudes from all the seismograms are collected and a least squares inversion is done (Eq. 3.21), resulting in estimates of the model parameters at each two-way-time sample (which can be converted to depth from the knowledge of the velocity). Thus, an AVO inversion essentially maps the seismic traces into sections of fractional changes in elastic parameters or the AVO intercept and gradient. Note, however, that such an inversion leaves the imprint of the wavelet on the resulting sections, because the inversion is typically carried out on the standard NMO-corrected gathers that have been processed for true amplitude.

An example of AVO inversion for a 2D line is shown in **Fig. 5.2** (from Xia *et al.* 1998). Eq. 5.1 is used here as the forward modeling operator and was applied to data from a 2D seismic line in which CMP gathers were analyzed in the plane-wave domain. The density was assumed to be known, and, therefore, only the parameters

$$\frac{\Delta\alpha}{\alpha} \text{ and } \frac{\Delta\beta}{\beta}$$

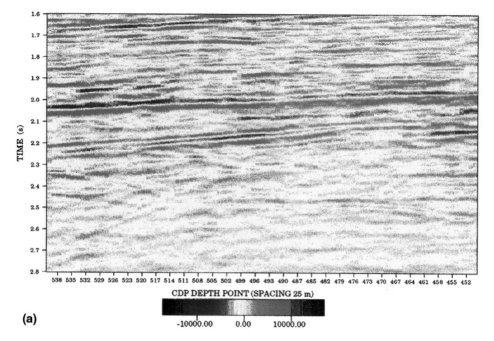

TIME (s)

CDP DEPTH POINT (SPACING 25 m)

-10000.00   0.00   10000.00

(a)

Fig. 5.2a—A 2D stacked section of a line: AVO analysis was carried out on all the CMP gathers along this line.

were estimated. The

$$\frac{\Delta\alpha}{\alpha} \text{ and } \frac{\Delta\beta}{\beta}$$

sections derived from the seismic line are shown in Figs. 5.2b and c, respectively. Most commonly, the AVO intercept and gradient terms are estimated from AVO analysis. The meanings and interpretation of the A and B terms (described in the following section) are very similar to the

$$\frac{\Delta\alpha}{\alpha} \text{ and } \frac{\Delta\beta}{\beta}$$

terms derived by Xia *et al.* (1998). Numerous examples of application of two-term AVO can be found in the literature. In fact, AVO analysis is now routinely done for hydrocarbon exploration.

5.1.3 AVO Inversion by Waveform Fitting. In an AVO waveform inversion approach, the forward modeling comprises the generation of synthetic seismograms at different offsets or angles using a convolution model similar to the one used in post-stack modeling (Chapter 2, Eq. 2.33), except that a two-term AVO equation is now used to compute synthetic seis-

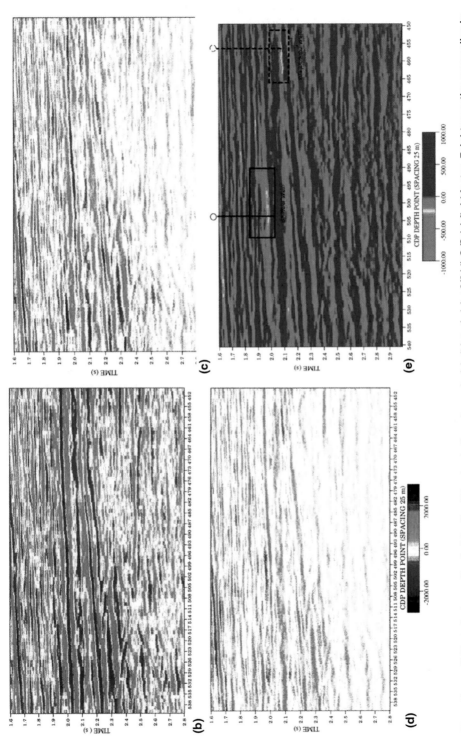

Fig. 5.2b–e—AVO attributes estimated for the 2D seismic line shown in (a). (b) $\Delta\alpha/\alpha$ (c) $\Delta\beta/\beta$ (d) Q/Q (e) fluid factor. Bright negative anomalies in the fluid factor section indicate possible hydrocarbon occurrence. Solid rectangle in (e) denotes a confirmed gas reservoir and the dashed rectangle in (e) is the recommended drilling site (from Xia *et al.* 1998).

mograms at a given angle or offset. The model parameters are generally estimated by iterative fitting of synthetic seismograms with the field data. The model parameter estimation problem is cast as a quasilinear inverse problem and an optimal model is generally estimated using a conjugate gradient method. The use of global optimization methods such as SA and GA has also been proposed, but a local optimization method seems appropriate for this application. Although the waveform inversion approach still uses the NMO-corrected gathers (assuming that the NMO or the background velocity is known), it attempts to fit the entire set of seismograms (of all angles and all arrivals) by iteratively perturbing a set of model parameters with proper constraints. Unlike the method described in the previous section, the waveform inversion approach makes use of a wavelet estimated from post-stack data in the computation of synthetic seismograms. Therefore, the resulting model parameter sections do not carry any imprint of the wavelet. Commercial software packages are now available and are routinely used in seismic data analysis. The first method, based on amplitude fitting, is very fast, and can be completed in about the same computation time as that of a conventional stack. On the other hand, the AVO waveform inversion method is expected to be more accurate.

5.1.4 AVO Interpretation. Identification of characteristic behavior of AVO curves for different sand types and lithology prompted rapid developments of techniques for AVO inversion. The interpretation of the results from AVO inversion (the AVO attributes) is, however, not trivial. There is significant overlap in elastic properties between different rock types, and therefore, *a priori* information in terms of well logs and petrophysics plays a very important role in the interpretation.

The first attempt to derive a meaningful interpretation of AVO attributes is credited to Smith and Gidlow (1987). Note that the amplitudes in seismograms—even after true amplitude processing—can be expected only to preserve relative amplitude. That is, the processed data can be calibrated only to reflection coefficients by matching with synthetic seismograms at well locations. For the inversion of AVO data and interpretation of AVO results, Smith and Gidlow (1987) proposed the use of a relationship between α and β. The most commonly used relationship for water-saturated clastic silicate rocks is the *mud/rock line* of Castagna *et al.* (1985), which is given by

$$\alpha = 1.36 + 1.16\beta, \quad \dots\dots\dots\dots\dots\dots\dots\dots\dots\dots\dots\dots\dots\dots (5.4)$$

where the velocities are in km/s. Such relationships can also be derived from well-log data for application in specific areas.

Having derived the AVO attribute sections such as

$$\frac{\Delta\alpha}{\alpha} \text{ and } \frac{\Delta\beta}{\beta}$$

(Fig. 5.2b and 5.2c), we can also derive two other sections called the *interpreter's sections* (Smith and Gidlow 1987) to highlight the fluid-content information. The principle behind generating the interpreter's section is based on the assumption that some deviation from the background trend may be attributed to fluid saturation. One such parameter is Poisson's ratio; it is generally more convenient to consider the ratio of

$$\frac{\alpha}{\beta},$$

which is related to Poisson's ratio by

$$\frac{\alpha}{\beta} = \left(\frac{1 - \sigma}{\frac{1}{2} - \sigma}\right)^{\frac{1}{2}} = Q. \quad \dots\dots\dots\dots\dots\dots\dots\dots\dots\dots\dots\dots (5.5)$$

Note that the ratio

$$\frac{\alpha}{\beta}$$

increases with an increase of Q. A pseudo-Poisson's ratio reflectivity is defined to reflect changes in Poisson's ratio by the following equation:

$$\frac{\Delta Q}{Q} = \frac{\Delta\alpha}{\alpha} - \frac{\Delta\beta}{\beta}. \quad \dots\dots\dots\dots\dots\dots\dots\dots\dots\dots\dots\dots (5.6)$$

Recall that the P-wave velocity generally decreases across an interface between the water-bearing and gas-bearing sandstones, while the S-wave velocity remains almost the same. Therefore, the Poisson's ratio and Q decrease at the top of a gas-filled sand and increase at the bottom. The relative contrast in Q can be estimated directly from the two AVO attribute sections.

A *fluid-factor reflectivity* can be defined to search for any deviation from an assumed relationship between α and β, which for the mud/rock line becomes

$$\Delta F = \frac{\Delta\alpha}{\alpha} - 1.16\frac{\beta}{\alpha}\frac{\Delta\beta}{\beta}. \quad \dots\dots\dots\dots\dots\dots\dots\dots\dots\dots\dots\dots (5.7)$$

Note that the fluid factor is zero if α and β follow the assumed background trend; nonzero ΔF indicates any deviation from the trend. Because the gas-bearing sandstone separates nicely from the water-bearing sandstone, ΔF should be negative at the top of a gas-filled sand and positive at the base.

The parameters pseudo-Poisson's ratio and the fluid factor are determined completely from the data and the assumed functional relationship between α and β. For the data example shown in Fig. 5.2a, the pseudo-Poisson's ratio and the fluid-factor sections are shown in Figs. 5.2d and e, respectively. The zones of negative

$$\frac{\Delta Q}{Q}$$

and ΔF are clearly observable in these two sections. Two predominant gas reservoirs are obvious, as indicated by the negative values of ΔF. Detailed comparison between well data and the inverted fluid factor was done at one location (Fig. 5.2f). Note the strong negative

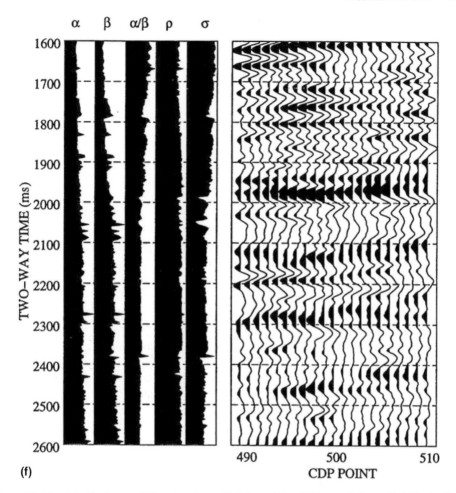

(f)

Fig. 5.2f—The left displays well-log data from Well A, and the right panel shows fluid factor for 20 CMP locations centered around Well A. The strong negative anomalies at about 1.97 seconds correlate with the well data very well (from Xia *et al.* 1998).

amplitude in the ΔF profile at slightly before 2.0 seconds, which correlates closely with the well data.

Following the Smith and Gidlow (1987) paper, several investigations were reported in which AVO attributes such as A and B and some combinations thereof were crossplotted for fluid identification. The basic principle behind this approach is that under a variety of petrophysical interpretations, brine-saturated sandstone and shale follow a background trend in the A-B plane. Foster *et al.* (1997) interpreted the A vs. B crossplot with the help of reference points such as *fluid angle* or *fluid line*.

Castagna *et al.* (1998) presented a systematic approach to the interpretation of AVO gradient and intercepts. They showed that with realistic petrophysical assumptions, equations for the *background trend* or *fluid line* could be derived resulting in a simple interpretation of crossplots. They demonstrated that under such assumptions, nonhydrocarbon-bearing clastic rocks often exhibit a background trend in the *A-B* plane (**Fig. 5.3**); *A* and *B* are found to be negatively correlated at very high

$$\frac{\alpha}{\beta}$$

ratio for background rocks, but they may be positively correlated at very high

$$\frac{\alpha}{\beta}$$

ratios in very soft shallow sediments. Thus, hydrocarbon occurrence can be inferred from variations away from the trend. Castagna *et al.* (1998) reevaluated the three AVO classes proposed by Rutherford and Williams (1989) in the context of crossplots and included an additional Class IV type AVO. Contrary to the common assumption that gas sand amplitude increases with offset, Castagna *et al.* (1998) demonstrated that gas sands may exhibit a wide variety of AVO behavior. Their findings are summarized in Fig. 5.3 and **Table 5.1**, where we can identify the location of different AVO types in the *A-B* plane. Results such as these are crucial to the meaningful interpretation of AVO attributes. One should, however, care-

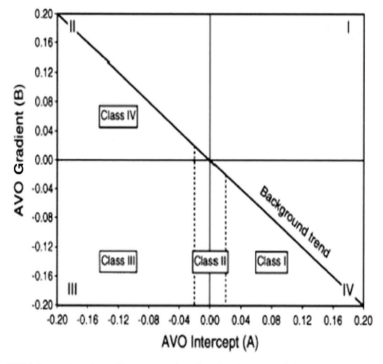

Fig. 5.3—AVO intercept and gradient crossplot showing four possible quadrants. For a limited time window, brine-saturated sandstones and shales tend to fall along a well-defined background trend. Top of gas/sand reflections tends to fall below the background trend, whereas the base of gas/sand reflections tends to fall above the trend (from Castagna *et al.* 1998; reprinted with permission from the Soc. of Exploration Geophysicists).

TABLE 5.1—TOP GAS-SAND REFLECTION COEFFICIENT VS. OFFSET BEHAVIOR FOR THE THREE RUTHERFORD AND WILLIAMS (1989) CLASSES I–III ASSUMING A TYPICAL "BACKGROUND" TREND WITH NEGATIVE SLOPE*

Class	Relative Impedance	Quadrant	A	B	Remarks
I	Higher than overlying unit	IV	+	−	Reflection coefficient (and magnitude) decrease with increasing offset
II	About the same as the overlying unit	III or IV	±	−	Reflection magnitude may increase or decrease with offset, and may reverse polarity
III	Lower than overlying unit	III	−	−	Reflection magnitude increases with offset
IV	Lower than overlying unit	II	−	+	Reflection magnitude decreases with offset

*Class IV sands, though not explicitly discussed by Rutherford and Williams, may be considered to be a subdivision of their Class III sands (from Castagna *et al.* 1998).

fully check the validity of the assumptions involved in deriving these "rules of thumb" prior to application to data from a new area.

Other approaches to AVO interpretation include crossplotting of various combinations of the Lamé parameters λ and μ.

5.1.5 AVO for Carbonate Rocks. All of the discussions on AVO in the preceding sections concerned clastic reservoirs. Chako (1989) reported on the use of AVO in the south Sumatra basin. He noted that the P-wave reflection coefficients from tops of limestone beds are usually lower where porosity exists, and, therefore, such amplitude anomalies may indicate porous zones. However, such information may not distinguish porosity from facies changes. Chako (1989) showed that AVO modeling can indeed distinguish between porous and tight limestone facies in his study area, because they show unique and different reflection coefficient variation with offset.

5.1.6 Limitations of AVO. The AVO analysis assumes that the amplitude recorded by a time sample in a seismic trace is caused by reflection from a plane interface. In other words, it assumes a "P-wave primaries only" reflection model from a welded contact of two elastic half spaces, such that the amplitude variations can only be related to angle-dependent plane-wave reflection coefficients. In Chapter 2, we listed several factors that affect seismic amplitudes, of which reflection coefficient is one. This simply means that either the effects of all the factors other than the reflection coefficients have been removed from the data or other effects are negligible. In practice, some of the effects can indeed be minimized, but several factors may still affect the amplitude. They are described below.

- Incorrect NMO velocities—Residual NMO-AVO attributes are estimated by weighted stacking of amplitudes from NMO-corrected gathers. Incorrect NMO velocities will result in either under- or overcorrection, and, thus, incorrect amplitudes will be used in the AVO inversion.

- Anisotropy—Common AVO analysis makes use of reflection coefficients at an interface from two isotropic elastic layers. Anisotropy, if present (because of shales and layering in the sediments), generally causes nonhyperbolic moveout that cannot be corrected by using simple NMO equations. In addition, the reflection coefficients for an interface between anisotropic shale and isotropic sand show behavior that is distinctly different from that of isotropic reflection coefficients.
- Lateral heterogeneity—AVO assumes a locally-1D Earth model, and the presence of strong lateral heterogeneity is certain to cause AVO analysis to fail. Generally, the first-order effects of lateral heterogeneity are corrected by time-migration of seismic data prior to AVO analysis.
- Thin layering—Reflection amplitudes from thin layers include composite effects of reflections from a zone, including all internal multiples and converted waves.
- Overburden effects—AVO assumes a single interface model. However, in a stack of layers, transmission through the overlying layers may have a significant effect on AVO.
- Geometrical spreading—Spreading corrections are generally computed using approximate formulas for 1D models, and they can easily break down in complex geologic regions.

5.2 Elastic Impedance
The equation for the plane-wave reflection coefficient for nonnormal incidence is much more complicated than that for the normal-incidence reflection coefficients. Shuey's two-term and three-term expressions are easier to interpret than the exact Zoeppritz coefficients in that the intercept term represents the normal-incidence reflection coefficients. The interpretation of the AVO gradient term is not very clear, and in general, it is not very straightforward to calibrate. Unlike the post-stack inversion that essentially inverts an entire trace using a model-based inversion, AVO waveform inversion is not widely practiced. Connolly (1999) derived an elegant expression for a non-normal-incidence reflection coefficient. It has the same form as a normal-incidence reflection coefficient, which enables one to implement the sparse-spike-type post-stack inversion algorithms to prestack angle gathers. In addition, it offers the potential for better calibration of results from inversion of nonnormal angle gathers. Analogous to the expression for the normal-incidence reflection coefficient given by Eq. 2.30, Connolly (1999) defined the nonnormal-incidence reflection coefficient as follows:

$$R_{pp}^i(\theta) = \frac{E_{i+1} - E_i}{E_{i+1} + E_i}, \quad \dots\dots\dots\dots\dots\dots\dots\dots\dots\dots\dots\dots\dots\dots (5.8)$$

where $R_{pp}^i(\theta)$ is the plane P-wave reflection coefficient at an angle θ for an interface i, and E_i represents *elastic impedance* (EI) for layer i. Connolly carried out the algebra to convert the reflection coefficient given by Eq. 2.27 into the form of Eq. 5.8 under a weak contrast approximation. The EI is given by

$$E_i = \alpha_i^{1+\tan^2\theta} \beta_i^{-8K\sin^2\theta} \rho_i^{(1-4K\sin^2\theta)}, \quad \dots\dots\dots\dots\dots\dots\dots\dots\dots\dots\dots (5.9)$$

with

$$K = \frac{\left[\left(\dfrac{\beta_i}{\alpha_i}\right)^2 + \left(\dfrac{\beta_{i+1}}{\alpha_{i+1}}\right)^2\right]}{2}. \qquad\ldots\ldots\ldots\ldots\ldots\ldots\ldots\ldots\ldots\ldots\ldots\ldots\ldots (5.10)$$

Dropping the third term in Eq. 2.27 is equivalent to replacing the $\tan^2\theta$ term in Eq. 2.27 with $\sin^2\theta$ in the Connolly formula (Eq. 5.9) resulting in

$$E_i = \alpha_i^{1+\sin^2\theta}\beta_i^{-8K\sin^2\theta}\rho_i^{(1-4K\sin^2\theta)}. \qquad\ldots\ldots\ldots\ldots\ldots\ldots\ldots\ldots\ldots\ldots\ldots\ldots (5.11)$$

The motivation behind EI formulation is demonstrated in **Fig. 5.4**; it shows that a 30° EI log is broadly similar in appearance to the AI log, allowing for calibration with wells. **Fig. 5.5** also demonstrates that large-angle EI gives better discrimination at the top of the gas reservoir (Veeken and De Silva 2004).

The salient features of EI inversion are summarized below:

- The reflection coefficient at nonnormal incidence has the same form as that at normal incidence.
- Post-stack inversion algorithms can be applied to angle gathers to derive EI models.
- The cost of EI inversion is very low, and the procedure is similar to that of an AI inversion. Wavelets are derived for different incident angle traces, and the time-migrated traces are inverted for EI, which are calibrated at the well locations.

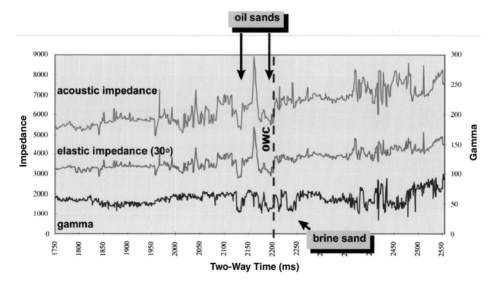

Fig. 5.4—Comparison of an AI curve with a 30° EI curve. Notice the broad similarity between the two (from Connolly 1999; reprinted with permission from the Soc. of Exploration Geophysicists).

Elastic impedance at different offset angles

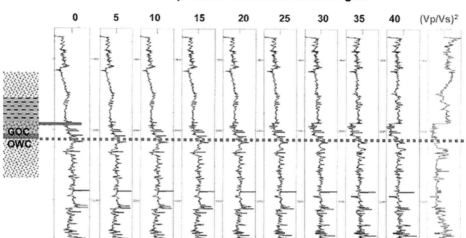

Fig. 5.5—EI logs computed for different angles: Note that the top of the gas reservoir is well expressed at large angles (from Veeken and De Silva 2004).

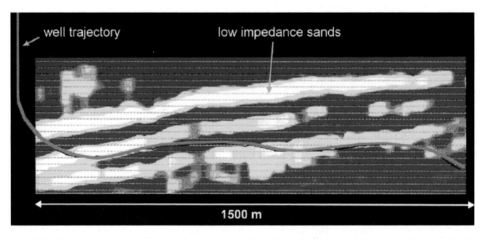

Fig. 5.6—A 30° EI volume estimated from seismic data: The figure shows the path of the first development well. The locations of oil-bearing sands encountered by the well correlates with the areas of low EI (from Connolly 1999; reprinted with permission from the Soc. of Exploration Geophysicists).

- The elastic model is extracted from a linear fit to the logarithm of the EI, and thus several interpreters' volumes such as AI, Poisson's ratio, and Lamé parameters can be derived. Crossplots of different derived parameters can be used in the interpretation of lithology and fluid. An example of EI inversion is shown in **Figs. 5.7 and 5.8** (from Lu and McMechan 2002).
- Although based on approximations, such an approach generally estimates P- and S-impedance reasonably well.

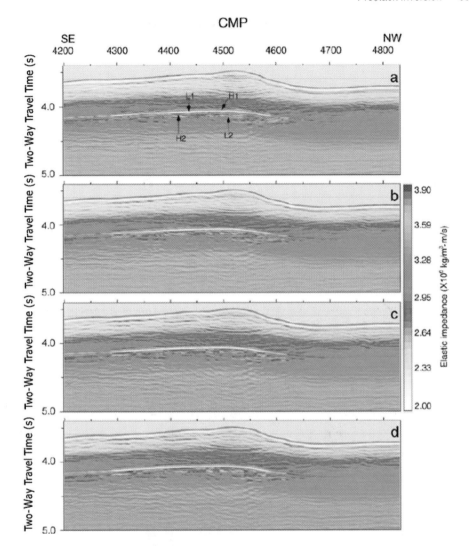

Fig. 5.7—An example of EI inversion of a 2D line: The figure shows EI sections at angles of (a) 0°–8°, (b) 8°–16°, (c) 16°–24°, and (d) 24°–32°. The L1 and L2 are low-EI layers and H1 and H2 are high EI layers (from Lu and McMechan 2002; reprinted with permission from the Soc. of Exploration Geophysicists).

5.3 Full Waveform Inversion

The motivations for the development of prestack full waveform inversion algorithms are demonstrated in **Fig. 5.9**. We blocked a well log at two different blocking factors (coarse and fine) and computed synthetic seismograms using two algorithms, namely, a convolution modeling and a reflectivity approach. The convolutional modeling algorithm uses exact reflection coefficients for seismic amplitude modeling and thus includes primary P-wave reflections only. On the other hand, the reflectivity algorithm described in Chapter 2 includes all the internal multiples, transmission losses, and converted waves. Note that

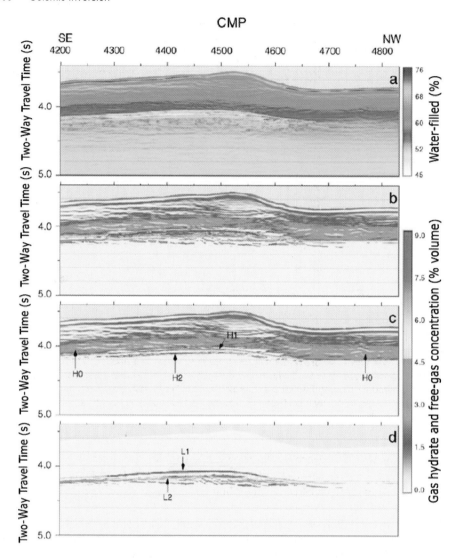

Fig. 5.8—Interpretation of results from EI inversion: (a) water-filled porosity, (b) gas hydrate and free gas concentration, (c) gas hydrate concentration, and (d) free gas concentration (from Lu and McMechan 2002; reprinted with permission from the Soc. of Exploration Geophysicists).

the AVO and EI algorithm described in the preceding sections are based on "P-wave primaries only" models, while the reflectivity modeling is more exact for locally 1D Earth models. The synthetic seismograms generated by the two methods were corrected for NMO and spreading, then converted to angle gathers. The processed seismograms are displayed in Fig. 5.9 together with their corresponding models. Note that when the well log is approximated with thick layers (isolated interfaces), the two sets of seismograms are fairly similar. However, when the same well log is approximated with many thin layers, we notice substantial differences between the two sets of seismograms; the differences become

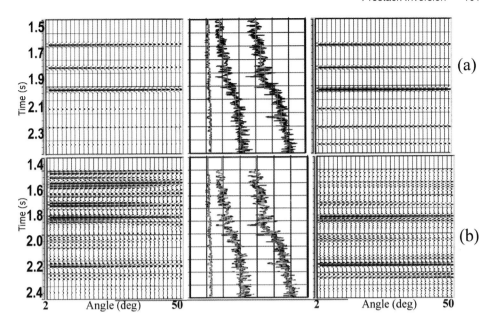

Fig. 5.9—Comparison between convolutional (P-wave primaries only) modeling and full wave-form modeling: A well log was blocked with different resolutions for computing synthetic seismograms. (a) Well log was approximated with a few thick layers. The P-wave primaries, only synthetics (left panel), compares well with those computed by full wave formulation. (b) Well log was approximated with many fine layers (middle panel). The P-wave primaries, only synthetics (left panel), differ considerably from those computed with full wave formulation, especially at angles larger than 20°.

larger as the angles of incidence increase. We can, therefore, clearly appreciate the benefit of full waveform seismogram calculation in modeling field data.

A full waveform prestack inversion algorithm makes use of an accurate modeling algorithm such as the reflectivity approach for the generation of synthetic seismograms, and a model-based inversion approach (see Chapter 3) for finding an optimal model described by the P-wave velocity, S-wave velocity, density, and thickness (or the P-wave velocity, impedance, and Poisson's ratio) of different layers. Because the forward modeling using reflectivity is computationally very expensive, one needs to make careful choices for the following factors:

- *Data Domain*: Generation of synthetic seismograms in the offset-time domain by the reflectivity method requires that plane-wave seismograms be evaluated first. These are then converted to the offset-time domain using the Fourier-Hankel transform. The procedure, however, requires the calculation of numerous plane-wave seismograms to avoid aliasing in the offset-time domain. The plane-wave or τ-p seismograms and the angle domain seismograms can easily be generated from the frequency domain plane-wave responses. Often the data are transformed to the τ-p or angle domain; this completely avoids conversion to the offset-time domain, resulting in a very fast algorithm. In addition, a true amplitude plane-wave transformation of data automatically corrects the data for spreading.

- *Optimization Method*: Unlike the AVO or EI analysis, a prestack waveform inversion generally makes use of CMP gathers that have not been NMO-corrected. This results in a nonlinear problem, and, therefore, a global optimization method (Sen and Stoffa 1995) may be an ideal choice. For a large number of optimization problems, global optimization methods may be too computer-intensive. With proper constraints, local optimization methods (Tarantola 1987) can also be employed. They, however, require that the differential seismograms (derivative of data with respect to model parameters) be evaluated. Sen and Roy (2003) outlined a semianalytic method for differential seismogram calculation that is computationally very efficient. In practice, a hybrid algorithm that takes advantage of both the local and global optimization methods is often found to be practical for application to large data sets.

- *Objective function and* a priori *constraints*: The standard L_2 norm is often employed in prestack waveform inversion. The choice of the model norm is, however, crucial because of the nonuniqueness of the inverse problem. Low-frequency trends in S-wave velocity and density are crucial to estimating realistic solutions, and a realistic low-frequency P-wave velocity trend is important if local optimization methods are used to avoid getting trapped in local minima. For global optimization algorithms, it is important to choose realistic bounds for the model parameters.

Fig. 5.10 shows the results from prestack waveform inversion of a CMP gather from a 2D line from the Gulf of Thailand (Roy *et al.* 2004). The inversion was carried out with a gradient-descent-based optimization method with iteration-adaptive regularization weight. Low-frequency trends were derived from the well logs and interactive interval velocity analysis, and data were inverted in the τ-p domain. The maximum angle used in the analysis was 54.8°. The upper panel (Fig. 5.10a) shows the estimated model parameters and displays the level of detail that is observable in such results. A zone of low impedance and low Poisson's ratio is clearly observable. All the different interpreters' attributes can be generated from these results. Fig. 5.10b shows the NMO-corrected gather together with the best-fit synthetic seismograms and the data residual; the data fit is excellent. **Fig. 5.11a** shows the 2D line and Fig. 5.11b shows the Poisson's ratio section derived by inverting all the CMP gathers along the 2D line. A zone of low Poisson's ratio is clearly visible in the 2D section. Detailed interpretation of the results is presented in Roy *et al.* (2004).

Fig. 5.12 displays impedance sections estimated by post-stack and prestack inversion algorithms. The prestack inversion used here is a hybrid algorithm that is a combination of a genetic algorithm and AVO (Mallick 2001). Notice the improved resolution in the impedance section obtained by prestack waveform inversion. This example clearly demonstrates the superiority of prestack inversion even in the estimation of acoustic impedances. As noted in the preceding paragraph, prestack inversion also enables us to estimate Poisson's ratio—an important attribute useful for hydrocarbon detection. **Fig. 5.13** displays the results from a hybrid prestack waveform inversion of a 3D seismic dataset from the Maria Ines Oste field in Argentina (Benabentos *et al.* 2002): The top panel displays the amplitude slice over a portion of the seismic data and the Poisson's ratio estimated from waveform inversion. The differences between oil and gas discoveries are not discernible in seismic data, but the two are distinct in the Poisson's ratio section. The bottom panel shows the pseudo-Poisson's ratio derived from AVO in the left column and the waveform-inversion-derived Poisson's ratio in the right column. Unlike the AVO derived Poisson's ratio section, the

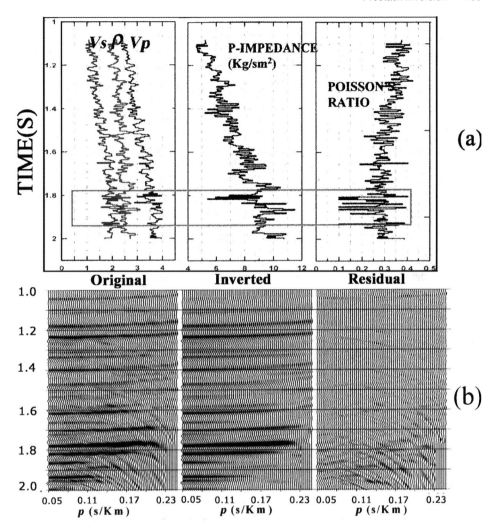

Fig. 5.10—Results from full wave pre-stack constrained waveform inversion of a CMP gather: (a) Derived elastic parameter profiles, (b) left panel shows the NMO-corrected data, the middle panel shows modeled data and the right panel shows the residual (modified from Roy *et al.* 2004; reprinted with permission from the Soc. of Exploration Geophysicists).

waveform inversion clearly distinguishes between oil and gas reservoirs. Note that the full waveform inversion derived distinctly different Poisson's ratios for the oil and gas reservoirs. Recall that the typical Poisson's ratio for shale is 0.25–0.3, and for oil-bearing and gas-bearing sands is 0.2–0.25 and 0.1–0.18, respectively (Mavko *et al.* 1998). This example clearly demonstrates the superiority of waveform inversion over AVO inversion.

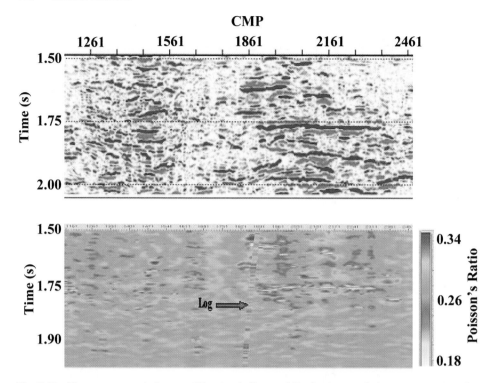

Fig. 5.11—The upper panel shows a 2D seismic line and the lower panel shows a cross section of Poisson's ratio section derived by a full waveform constrained prestack inversion (modified from Roy *et al.* 2004; reprinted with permission from the Soc. of Exploration Geophysicists).

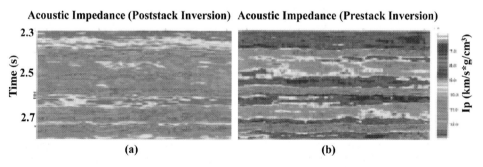

Fig. 5.12—Impedance section estimated from (a) post-stack and (b) pre-stack inversion of Woodbine gas sand (modified from Mallick 2001; reprinted with permission from the Canadian Soc. of Exploration Geophysicists).

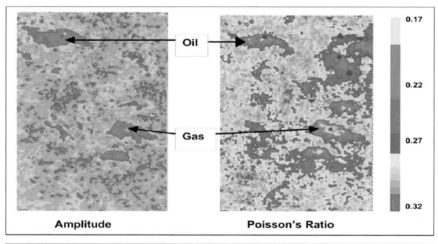

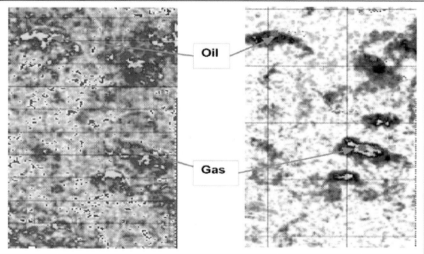

Fig. 5.13—Results from a hybrid prestack waveform inversion of a 3D seismic dataset from the Maria Ines Oste field, Argentina (Benabentos *et al.* 2002). The top panel displays amplitude slice over a portion of the seismic data and the Poisson's ratio estimated from waveform inversion. The differences between oil and gas discoveries are not discernible in seismic data, but the two are distinct in the Poisson's ratio section. The bottom panel shows the pseudo-Poisson's ratio derived from AVO in the left column and the waveform-inversion-derived Poisson's ratio in the right column. Unlike the AVO-derived Poisson's ratio section, the waveform inversion clearly distinguishes between oil and gas reservoirs (from Benabentos *et al.* 2002; reprinted with permission from the Soc. of Exploration Geophysicists).

Chapter 6

Summary

In the preceding chapters, we described the fundamentals of seismic wave propagation, inverse theory, and some of the most popular approaches to seismic inversion. Since the advent of bright spots, seismic amplitude information has been used routinely to obtain lithologic information and for the detection of hydrocarbons. Results from seismic inversion are now used extensively in building reservoir models. Of all the seismic waveform inversion methods, the post-stack inversion is the most commonly used. Although this approach is based on a simplistic wave propagation model (normal incidence), the algorithm is computationally efficient, and, therefore, 3D post-stack volumes can be processed rapidly. Meaningful results can be obtained when well-log information is available at a sufficient number of locations. For prestack analysis, AVO and recently developed EI inversion are the most popular, because AVO and EI inversion are only slightly more expensive than a post-stack inversion. Although theoretically more rigorous, full waveform inversion is not commonly used because of high computation cost. Much effort is directed toward the interpretation of results from inversion. Although the development of rigorous inversion algorithms is itself an important area of research, accurate interpretation of the results from inversion is the most important step toward building realistic reservoir models. As computers become faster and cheaper, more accurate methods will be employed in the foreseeable feature. Some of the more rigorous methods are already used in the industry on limited data sets, and others are in the development stage. We outline some of these methods below:

- *Resolution*: The resolution of results from seismic inversion is limited to the seismic frequency band, which is significantly coarser than the resolution of well logs. In a geostatistical inversion, higher frequencies are introduced into the seismic inversion results from well-log-derived statistical information such that the seismic data fit is not degraded. In some field data applications, post-stack geostatistical inversion has shown promising results. The extension of this method to prestack inversion is still under development.
- *Lateral heterogeneity*: The AVO, EI, and even full waveform inversion described in the previous chapters assume a locally 1D Earth model and are applied to CMP gathers. Although time migration is often applied to the seismic data prior to seismic inversion, all these methods fail in strongly heterogeneous media. Some attempts have

been made to make use of 2D and 3D elastic Earth models using finite differences, but these methods are still far from being practical. Excellent theoretical developments have been made in this area, and we hope that such methods will be used more extensively as computers get faster.

- *Attenuation*: The effect of attenuation is generally ignored in seismic inversion. However, in many situations such as in gas-charged sediments, amplitude loss because of attenuation is too large to be ignored. Although it is straightforward to include the effect of attenuation in the waveform inversion algorithms, it is extremely difficult to incorporate the effects of attenuation in AVO and EI inversion. If reasonable attenuation models become available, the amplitude data can be corrected for the effects of attenuation prior to AVO and EI.

- *Anisotropy*: In many geologic situations, the effect of anisotropy cannot be ignored. For example, in many areas, the shales show noticeable variation in seismic wave velocity with direction. Techniques are being developed to correct for the effects of anisotropy on AVO. Over fractured anisotropic reservoirs, the amplitude is reported to vary not only with offset but also with azimuth (AVOA), and AVOA is now used to estimate fracture density and orientation.

- *Uncertainty estimation*: To document the nonuniqueness of seismic inversion results, rigorous estimates of uncertainty are needed. Bayesian stochastic inversion methods can be employed for this purpose. Although it is not a common practice today because of limitations on computation, uncertainty estimates are useful in reservoir model-building, prediction, and planning.

Nomenclature

$\mathbf{C}$	=	fourth order elastic coefficient tensor
$\mathbf{C}_d$	=	data covariance matrix
$\mathbf{C}_m$	=	model covariance matrix
$\mathbf{d}$	=	data vector
$<\mathbf{d}>$	=	mean data vector
$\mathbf{e}$	=	error vector
$\mathbf{e}_d$	=	data error
$\mathbf{e}_m$	=	model error
E	=	Elastic Impedance
$\mathbf{G}^g$	=	generalized inverse
$\mathbf{G}^T$	=	matrix $\mathbf{G}$ transpose
$\mathbf{k}$	=	wavenumber vector
K	=	bulk modulus
$\mathbf{m}$	=	model vector
$\hat{\mathbf{o}}$	=	second order stress tensor
p	=	horizontal slowness
$\mathbf{p}$	=	slowness vector
q	=	vertical slowness
R_{pp}	=	P-P reflection coefficient
R_{ps}	=	P-S reflection coefficient
t	=	time
T	=	temperature
Z	=	Acoustic Impedance
$\|\cdot\|$	=	norm
α	=	compressional wave velocity
β	=	shear-wave velocity
$\mathring{\mathbf{a}}$	=	second order strain tensor
λ, μ	=	Lamé parameters
ρ	=	density
θ	=	angle of incidence
τ	=	vertical delay time
ϕ	=	porosity
ω	=	angular frequency

References

Aki, K., and Richards, P.G. 2002. *Quantitative Seismology*, second edition, Sausalito, California: University Science Books.

Backus, G.E. 1962. Long-wave elastic anisotropy produced by horizontal layering. *J. of Geophysical Research*, **67**:4427–4440.

Benabentos, M., Mallick, S., Sigimondi, M., and Soldo, J. 2002. Seismic reservoir description using hybrid seismic inversion: A 3D case study from the Maria Ines Oste field, Argentina. *The Leading Edge*, **21**(10):1002–1008.

Biot, M.A. 1956. Theory of propagation of elastic waves in fluid-saturated porous solid. *J. Acoust. Soc. Am.*, **26**:182–185.

Castagna, J.P. and Backus, M.M. 1993. *Offset dependent reflectivity—Theory and practice of AVO analysis*. Tulsa: Society of Exploration Geophysicists Press.

Castagna, J.P., Batzle, M.L., and Eastwood, R.L. 1985. Relationships between compressional and shear-wave velocities in clastic silicate rocks. *Geophysics*, **50**(4):571–581.

Castagna, J.P., Swan, H.W., and Foster, D.J. 1998. Framework for AVO gradient and intercept interpretation. *Geophysics*, **63**(3):948–956.

Cerveny, V. 2001. *Seismic Ray Theory*, New York/Cambridge: Cambridge University Press.

Chako, S. 1989. Porosity identification using amplitude variation with offset: Examples from south Sumatra. *Geophysics*, **54**(8):942–951.

Chunduru, R.K., Sen, M.K., and Stoffa, P.L. 1997. Hybrid optimization methods for geophysical inversion. *Geophysics*, **62**(4):1196–1207.

Connolly, P. 1999. Elastic impedance. *The Leading Edge*, **18**(4):438–452.

Cooke, D.A. and Schneider, W.A. 1983. Generalized linear inversion of reflection seismic data. *Geophysics*, **48**(6):665–676.

Dvorkin, J. and Alkhater, S. 2004. Pore fluid and porosity mapping from seismic data. *First Break*, **22**:53–57.

Foster, D.J., Keys, R.G., and Schmitt, D.P. 1997. Detecting subsurface hydrocarbons with elastic wavefields. In G. Chavent, G. Ooanicolaou, P. Sacks, and W. Symes (eds.), *Inverse Problems in Wave Propagation,* New York City: Springer-Verlag.

Gassmann, F. 1951. Elastic waves through a packing of spheres. *Geophysics*, **16**(3):673–685.

Ghosh, S.K. 2000. Limitations on impedance inversion of band-limited reflection data. *Geophysics*, **65**(3):951–957.

Gill, P.E., Murray, W., and Wright, M.H. 1981. *Practical Optimization*, San Diego, California/London: Academic Press.

Ingber, L. 1993. Simulated annealing: Practice versus theory. *J. of Mathematical and Computer Modeling,* **18**(11):29–57.

Kelly, K.R., Ward, R.W., Treitel, S., and Alford, R.M. 1976. Synthetic Seismograms: A finite-difference approach. *Geophysics,* **41**(1):2–27.

Kennett, B.L.N. 1983. *Seismic wave propagation in stratified media,* Cambridge/New York: Cambridge University Press.

Kirkpatrick, S.G., Gelatt, C.D. Jr., and Vecchi, M.P. 1983. Optimization by simulated annealing. *Science,* **220:**671–680.

Lay, T. and Wallace, T. 1995. *Modern Global Seismology,* San Diego, California: Academic Press.

Levander, A. 1988. Fourth-order finite-difference of P-SV seismograms. *Geophysics,* **53**(11):1425–1436.

Lu, S. and McMechan, G.A. 2002. Estimation of gas hydrate and free gas saturation, concentration, and distribution from seismic data. *Geophysics,* **67**(2):582–593.

Mallick, S. 2001. Prestack waveform inversion using a genetic algorithm: The present and the future. *CSEG Recorder,* 78–84.

Mavko, G., Mukerji, T., and Dvorkian, J. 1998. *Rock Physics Handbook: Tools for Seismic Analysis in Porous Media,* New York/Cambridge: Cambridge University Press.

Menke, W. 1984. *Geophysical Data Analysis: Discrete Inverse Theory,* Orlando, Florida: Academic Press.

Nur, A.M. and Wang, Z. 1998. *Seismic and acoustic velocities in reservoir rock,* Volume 1, Experimental Studies: Geophysics Reprint Series, Franklyn K. Levin (ed.), Tulsa: Society of Exploration Geophysicists.

O'Connell, R.J. and Budiansky, B. 1978. Measures of dissipation in viscoelastic media. *Geophysical Research Letters,* **5:**5–8.

Oldenburg, D.W., Levy, S., and Whitall, K.P. 1981. Wavelet estimation and deconvolution. *Geophysics,* **46**(11):1528–1542.

Ostrander, W.J. 1984. Plane-wave reflection coefficients for gas sands at non-normal angles of incidence. **49**(100):1637–1648.

Press, W.H., Flannery, B.P., Teukolsky, S.A., and Vetterling, W.T. 1992. *Numerical Recipes in C: The art of scientific computing,* New York/Cambridge: Cambridge University Press.

Ricker, N. 1953. The form and laws of propagation of seismic wavelets. *Geophysics,* **18**(1): 10–40.

Roy, I.G., Sen, M.K., Torres-Verdin, C., and Varela, O.J. 2004. Prestack inversion of a Gulf of Thailand OBC dataset. *Geophysics,* In Press.

Rutherford, S.R. and Williams, R.H. 1989. Amplitude-versus-offset in gas sands. *Geophysics,* **54**(6):680–688.

Sen, M.K. and Stoffa, P.L. 1995. *Global optimization methods in geophysical inversion,* The Netherlands: Elsevier Science Publications.

Sen, M.K. and Roy, I.G. 2003. Computation of differential seismograms and iterative adaptive regularization in prestack waveform inversion. *Geophysics,* **68**(6):2026–2039.

Sheriff, R.E. and Geldart, L.P. 1995. *Exploration Seismology,* New York/Cambridge: Cambridge University Press, 592 pages.

Shipboard Scientific Party, Leg 204. 2002. *Preliminary report: ODP Preliminary Report 204* [online]: http://wwwodp.tamu.edu/publications/prelim/204_prel/204PREL.PDF.

Shuey, R.T. 1985. A simplification of Zoeppritz equations. *Geophysics,* **50:**609–614.

Smith, G., and Gidlow, P.M. 1987. Weighted stacking for rock property estimation and detection of gas. *Geophysical Prospecting*, **35:**993–1014.

Smith, T.M., Sondergeld, C.H., and Rai, C.S. 2003. Gassmann fluid substitution: A tutorial. *Geophysics*, **68**(2):430–440.

Stolt, R.H. 1989. Inversion. Seismic Inversion Revisited. In *Geophysical Inversion*, B. Bedner, L.R. Lines, R.H. Stolt, and A.B. Weglein (eds.), Philadelphia, Pennsylvania: SIAM Press, 1–20.

Tarantola, A. 1987. *Inverse Problem Theory: Methods of Data Fitting and Model Parameter Estimation*, Investigations in Geophysics No. 6, Amsterdam/New York: Elsevier Science Publications.

Tatham, R.H. and McCormack, M.D. 1998. *Multicomponent Seismology in Petroleum Exploration*, Tulsa: Society of Exploration Geophysicists.

Telford, W.M., Geldart, L.P., and Sheriff, R.E. 1990. *Applied Geophysics*, New York/Cambridge: Cambridge University Press.

Thomsen, L. 1986. Weak Elastic Anisotropy. *Geophysics*, **51:**1954–1966.

Veeken, P.C.H. and De Silva, M. 2004. Seismic inversion methods and some of their constraints. *First Break*, **22:**47–70.

Wiggins, R.A. 1977. Minimum entropy deconvolution. *Proc., Intl. Symposium on Computer Aided Seismic Analysis and Discrimination*, Falmouth, Massachusetts: IEE Computer Society (9–10 June) 7–14.

Wood, W.T. 1999. Simultaneous deconvolution and wavelet inversion as a global optimization. *Geophysics*, **64**(4):1108–1115.

Wyllie, M.R.J., Gregory, A.R., and Gardner, G.H.F. 1958. An experimental investigation of factors affecting elastic wave velocities in porous media. *Geophysics*, **23**(3):459–493.

Xia, G., Sen, M.K., and Stoffa, P.L. 1998. 1-D elastic waveform inversion: a divide-and-conquer approach. *Geophysics*, **63**(5):1670–1684.

Yilmaz, O. 2001. *Seismic Data Analysis: Processing, Inversion and Interpretation of Seismic Data*, Tulsa: SEG Press.

Index